Astrovirus Research

Stacey Schultz-Cherry
Editor

Astrovirus Research

Essential Ideas, Everyday Impacts, Future Directions

Springer

Editor
Stacey Schultz-Cherry
Infectious Diseases
St Jude Children's Research Hospital
Memphis, Tennessee, USA

ISBN 978-1-4614-4734-4 ISBN 978-1-4614-4735-1 (eBook)
DOI 10.1007/978-1-4614-4735-1
Springer New York Heidelberg Dordrecht London

Library of Congress Control Number: 2012947401

Printed on acid-free paper

Springer is part of Springer Science+Business Media (www.springer.com)

This book is dedicated to the memory of Ernesto Méndez, who left us in November 2011. Ernesto was a pioneer in the study of astroviruses and he will be missed by the entire field. His early departure left many unfinished projects, and for those of us who had the good fortune to know him, left a space that will be impossible to fill. He will also be remembered for his happiness and devotion to science and life, for his endless patience, and for his passion for teaching.

Preface

Since the discovery of astroviruses in 1975, the astrovirus field has continued to make new and exciting discoveries. Although typically overshadowed by other enteric viruses like rotavirus and norovirus, astroviruses are increasingly recognized as a major source for outbreaks of gastroenteritis and diarrhea in animals, including humans and birds worldwide. In the past several years, there has been a rapid increase in the number of astroviruses detected across diverse species with increasing evidences that astroviruses likely cross species barrier and may be zoonotic. In spite of this increase in our knowledge, much work remains on understanding the basic virology, epidemiology, immunology, and pathogenesis of this varied group of viruses. One of the most pressing needs given the increasing viral diversity is a unified naming system that accounts for the species and genogroup of newly isolated viral strains. This and developing good diagnostic tools to understand the prevalence of astroviruses in nature will continue to move the field forward and increase our understanding of this fairly understudied, but interesting, group of viruses.

Memphis, TN, USA Stacey Schultz-Cherry

Acknowledgments

The authors would like to thank Mr. Michael Wilt, developmental editor, for his hard work and assistance. The book would not have happened without him. We also gratefully acknowledge all of our colleagues and laboratory members for their efforts through the years.

Contents

Contributors

Carlos F. Arias, Ph.D. Genética del Desarrollo y Fisiología Molecular, Instituto de Biotecnología, Universidad Nacional Autónoma de México, Cuernavaca, Mexico

Albert Bosch, Ph.D. Enteric Virus Laboratory, Department of Microbiology, Institute of Nutrition and Food Safety, University of Barcelona, Barcelona, Spain

Andrew Burnham, Ph.D. Department of Infectious Diseases, St. Jude Children's Research Hospital, Memphis, TN, USA

Giovanni Cattoli, D.V.M., Ph.D. Department of Research and Innovation, Istituto Zooprofilattico Sperimentale delle Venezie, Legnaro, Italy

Daniel Ka Wing Chu, Ph.D. Centre for Influenza Research and School of Public Health, The University of Hong Kong, Pokfulam, People's Republic of China

Kelly A. Dryden, Ph.D. Department of Molecular Physiology and Biological Physics, University of Virginia, Charlottesville, VA, USA

Rebecca M. Dubois, Ph.D. Department of Structural Biology, St. Jude Children's Research Hospital, Memphis, TN, USA

Stacy R. Finkbeiner, Ph.D. Department of Molecular Virology and Microbiology, Baylor College of Medicine, Houston, TX, USA

Susana Guix, Ph.D. Enteric Virus Laboratory, Department of Microbiology, Institute of Nutrition and Food Safety, University of Barcelona, Barcelona, Spain

Lori R. Holtz, M.D. Department of Pediatrics, Washington University School of Medicine/St. Louis Children's Hospital, St. Louis, MO, USA

Matthew D. Koci, Ph.D. Department of Poultry Science, North Carolina State University, Raleigh, NC, USA

Neel K. Krishna, Ph.D. Department of Microbiology and Molecular Cell Biology, Eastern Virginia Medical School, Norfolk, VA, USA

Victoria Meliopoulos, Ph.D. Department of Infectious Diseases, St. Jude Children's Research Hospital, Memphis, TN, USA

Rodrigo Velázquez, M.S. Genética del Desarrollo y Fisiología Molecular, Instituto de Biotecnología, Universidad Nacional Autónoma de México, Cuernavaca, Mexico

Andrea Murillo, M.Sc. Genética del Desarrollo y Fisiología Molecular, Instituto de Biotecnología, Universidad Nacional Autónoma de México, Cuernavaca, Mexico

Mary Pantin-Jackwood, D.V.M., M.S., Ph.D. Southeast Poultry Research Laboratory, United States Department of Agriculture, Agricultural Research Service, Athens, GA, USA

Malik Peiris, M.B.B.S., D.Phil., M.R.C.Path. Centre for Influenza Research and School of Public Health, The University of Hong Kong, Pokfulam, People's Republic of China

Rosa M. Pintó, Ph.D. Enteric Virus Laboratory, Department of Microbiology, Institute of Nutrition and Food Safety, University of Barcelona, Barcelona, Spain

Ernesto Méndez, Ph.D. Genética del Desarrollo y Fisiología Molecular, Instituto de Biotecnología, Universidad Nacional Autónoma de México, Cuernavaca, Mexico

Stacey Schultz-Cherry, Ph.D. Department of Infectious Diseases, St. Jude Children's Research Hospital, Memphis, TN, USA

Yizhi J. Tao, Ph.D. Department of Biochemistry and Cell Biology, Rice University, Houston, TX, USA

Daniel Todd, B.Sc., Ph.D. Virology Branch, Agri-Food and Biosciences Institute, Belfast, UK

Mark Yeager, M.D., Ph.D. Department of Molecular Physiology and Biological Physics, University of Virginia, Charlottesville, VA, USA

Chapter 1
Epidemiology of Human Astroviruses

Albert Bosch, Susana Guix, and Rosa M. Pintó

Abstract Molecular surveillance of viral gastroenteritis reveals that human astroviruses (HAstV) are one of the most important agents of acute pediatric diarrhea after rotaviruses and arguably caliciviruses. Sequence analysis of the capsid region enables the determination of HAstV genotype/serotype and their distribution into eight genotypes (HAstV-1–HAstV-8), with subtypes or lineages within the different genotypes. Additionally, two differentiated genogroups, A (containing HAstV-1–HAstV-5 and HAstV-8) and B (containing HAstV-6 and HAstV-7), have been defined based on the well-conserved partial sequence close to the protease motif coding region in the ORF1a. Worldwide, HAstV-1 is the most prevalent type.

In the recent years, metagenomic analysis of children's stool has led to the discovery of new HAstV strains highly divergent from the previously known genotypes and apparently related to animal astroviruses.

In temperate regions, HAstV infection occurs mostly in winter, although cases may be observed in summer, which may reflect a better stability of the virus at cold temperature. HAstV are transmitted through the fecal–oral route, and food and waterborne outbreaks have been reported in the literature. Moreover, fecally contaminated environmental surfaces act as vehicles for the transmission of the infection in day care centers, hospitals, nurseries, schools, and military institutions.

A. Bosch, Ph.D. (✉) • S. Guix, Ph.D. • R.M. Pintó, Ph.D.
Enteric Virus Laboratory and Institute of Nutrition and Food Safety,
Department of Microbiology, University of Barcelona,
Avenue Diagonal 643, Barcelona, 08028, Spain
e-mail: abosch@ub.edu; susanaguix@ub.edu; rpinto@ub.edu

S. Schultz-Cherry (ed.), *Astrovirus Research: Essential Ideas, Everyday Impacts, Future Directions*, DOI 10.1007/978-1-4614-4735-1_1,

Introduction

Madeley and Cosgrove coined in 1975 the term astrovirus to describe small round viruses with a distinct stellate appearance (*astron* means star in Greek) that were visualized by electron microscopy in the stool of children with diarrhea [69]. However, occurrence of the same 28–30 nm particles in stools of children with mild vomiting gastroenteritis had been previously reported by Appleton and Higgins [4].

Sequencing the astrovirus subgenomic RNA, Monroe and coworkers proposed the classification of astroviruses as a new family of RNA viruses, designated *Astroviridae* [81]. Human astroviruses (HAstV) belong to genus *Mamastrovirus* within this *Astroviridae* family [12] and will be the subject of this chapter.

Lee and Kurtz achieved in 1981 the trypsin-dependent passage of HAstV in cell monolayers [63] and set the procedures and materials (Oxford type-specific reference sera) to recognize at the time five serotypes of HAstV [57], which may be recognized as the birth of astrovirus epidemiology.

Burden of Disease Estimates

In a study on the annual burden of foodborne infections in the USA during the decade 1983–1992, Mead and colleagues estimated a total of 3,900,000 cases of astrovirus diarrhea per year [73]. Surveillance using molecular diagnostic methods developed in the past two decades shows that HAstV are one of the most important causes of pediatric acute gastroenteritis, after rotaviruses and arguably caliciviruses [39, 77, 114]. HAstV outbreaks affecting primarily young children have been reported worldwide [10, 88, 90, 104], although outbreaks affecting adults [8, 90] and the elderly [41, 64] are also frequently described. Additionally, HAstV are also the cause of sporadic cases and their incidence in children with gastroenteritis in both developing and developed countries ranges from 2% to 9% [10, 34, 37, 83, 103, 115], although some studies report much a higher prevalence [70]. HAstV have also been often associated with nosocomial infections in hospitals and they are reported to cause persistent diarrhea in immunocompromised hosts [101, 115]. They are often involved in outbreaks affecting bone marrow recipients [16, 18, 88] and patients from pediatric primary immunodeficiency units [38]. Of note is the increasing relevance of HAstV infections in patients with AIDS or other immune diseases [9, 42, 65, 118].

Although astrovirus surveillance studies were initially based on electron microscopy and enzyme immunoassay techniques, during the past few years the number of surveys using molecular techniques, mainly reverse transcription-polymerase chain reaction (RT-PCR), has substantially increased. There is a widespread belief that astrovirus incidence may have been underestimated since immunological techniques are far less sensitive than RT-PCR [39, 48].

Astrovirus Infection

In humans, astrovirus has been detected in duodenal biopsies in epithelial cells located in the lower part of villi [15]. Astrovirus gastroenteritis is usually a mild and self-limiting disease that does not require hospitalization. However, intussusception, a form of intestinal obstruction in which a segment of the bowel prolapses into a more distal segment has been observed in HAstV-infected infants [3]. Additionally, severe forms of astrovirus gastroenteritis have been reported [86, 90] and chronic diarrhea is often observed in immunocompromised individuals [18, 38]. Furthermore, a case of encephalitis has been reported in a 15-year-old boy with agammaglobulinemia [96]. Nevertheless, the high frequency of coinfections sometimes makes it difficult to evaluate the actual role and significance of HAstV infection in disease. Coinfections with other pathogens are detected from 33% to 65% of the astrovirus positive samples [17, 37, 47, 48, 115]; since some of these studies occurred in developing countries, it may be suggested that poor hygienic conditions contribute to multiple enteric infections.

Astrovirus infection is frequent in childhood and often unapparent. Seroprevalence studies indicate that most children acquire astrovirus antibodies early in life, with 7% of prevalence in 6–12-month-old infants and rising up to 75% in 10-year-old children [52, 54, 55].

Astrovirus Diversity

Astroviruses are RNA viruses and thus depend on error-prone polymerases that can contribute to a great deal of genetic variability [24]. Recombination events between structural and nonstructural coding regions further contribute to the genetic variability and phylogenetic complexity of astroviruses [7, 117].

On the basis of antigenic criteria, HAstV have been classified into eight serotypes (HAstV-1–HAstV-8) [12]. Although phylogenetic analysis of different parts of the genome results in different genetic clustering [7, 46, 68, 76, 100, 105], most studies based on the capsid region suggest a high correlation between serotypes and genotypes, a relationship that allows determining serotype information by sequence analysis. The comparison of sequences from the capsid region gives rise to radial tree topologies with equidistant clustering of serotypes. Moreover, based on the sequence diversity of the capsid region, subtypes [49, 66, 113] or lineages [7, 22, 36, 47, 74] may be established within the different genotypes.

Additionally, phylogenetic analysis of the well-conserved partial sequence close to the protease motif coding region results in only two clearly differentiated genogroups, called genogroup A (containing HAstV-1–HAstV-5 and HAstV-8) and genogroup B (containing HAstV-6 and HAstV-7) [7]. Moreover, genetic analysis of the hypervariable (HVR) region contained in the C-terminus of nsP1a coding region provides another typing method that enables identifying nine HVR genotypes within genogroup A and three within genogroup B [46].

The advent of the deep sequencing era has enabled studies at the metagenomic level. As an example, metagenomic analysis of the stool of a young boy with acute diarrhea in Melbourne, Australia, led to the discovery of a new species of HAstV, termed MLB1 that is highly divergent from all the previously known genotypes [29, 31]. Further work by the same laboratory demonstrated that human stool contains a previously unrecognized diversity of HAstV [30]. HAstV MLB1 has been thereafter found in diarrheic feces of children in the USA [32]. Similar approaches have led to the identification of other novel HAstV in patient stool as VA1, VA2, MLB2, and VA3, indicating the extreme divergence of these viruses [30]. Interestingly, astroviruses phylogenetically related with MLB1 have recently been detected in rat stool in Hong Kong, suggesting that they may share a common ancestor [20]. Kapoor and coworkers [50] also found novel astrovirus species (HMOAstV) in human stool related to mink and ovine astroviruses. One conclusion of their work is that each of the present three groups of astroviruses found in human stool (classic HAstV, AstV-MLB, and HMOAstV) is genetically more closely related to animal astroviruses (cats, rats, and sheep and minks, respectively) than to each other, indicating that human astroviruses may periodically emerge from zoonotic transmissions [50]. Recent reports suggest the occurrence of recombination events between human and animal strains [98, 108] and provide support to the belief that cross-species transmission of astrovirus has influenced their evolutionary history [111].

Astrovirus Prevalence and Distribution

Human astroviruses are distributed worldwide and have been generally associated with 2–9% of acute, nonbacterial gastroenteritis in children; although much higher incidences, up to 61% have been reported (Table 1.1 and reviewed in [21]). Overall, the mean worldwide incidence is around 11%, being 23% and 7% in rural and urban settings, respectively (Table 1.1). Mean incidences tend to be higher in developing countries, with the exception of sub-Saharan Africa, which shows low incidences of HAstV [6, 51, 97]. Interestingly and surprisingly, there is also a tendency of a reduction in the classical HAstV incidence worldwide in the last decades: 22% in the 1980s, 15% in the 1990s, and 5% in the 2000s (Table 1.1). This tendency is clearly seen, comparing data from surveillance studies performed in the same country 9 and 16 years apart, in France and Japan, respectively (Table 1.1). Further epidemiological studies are required to ascertain whether this observation actually reflects the present situation of classical astrovirus infections worldwide. What remains to be elucidated is whether AstV-MLB and HMOAstV strains outcompeting with classical HAstV strains may contribute to the decrease in the incidence of the classical strains, since surveillance and/or seroprevalence studies on the distribution of the new strains are not yet available.

In a large study conducted in Barcelona, Spain, the maximum classical astrovirus detection rate was observed in children between 2 and 4 years of age [47]. Reports in other countries such as Mexico, Thailand, Guatemala, France, Australia,

Table 1.1 Major prevalence studies based on classical astrovirus detection in stool

Study	Region (setting and period of study)	Age of study population	Detection method	Astrovirus prevalence (%)
North Africa				
Nafici et al. [86]	Egypt (rural, 1995–1998)	<2 years of age	EIA	26.7
Fodha et al. [33]	Tunisia (urban and rural 2003–2005)	<5 years of age	EIA	6.0
Sub-Saharan Africa				
Basu et al. [6]	Botswana (urban, 2001–2002)	<5 years of age	EIA	2.7
Reither et al. [97]	Ghana (urban, 2005–2006)	<12 years of age	RT-PCR	4.8
Kiulia et al. [51]	Kenya (urban and rural 1999–2005)	<5 years of age	EIA	5.3
Asia				
Herrmann et al. [48]	Thailand (urban, 1985–1987)	<5 years of age	EIA	8.6
Utagawa et al. [110]	Japan (urban and rural 1982–1992)	Children of all age groups. Mostly <10 years of age	EIA	18
Fang et al. [28]	China (urban and rural, 1998–2005)	<5 years of age	EIA plus RT-PCR	5.5
Nguyen et al. [87]	Vietnam (urban, 2005–2006)	<3 years of age	RT-PCR	13.9
Verma et al. [112]	India (urban, 2004–2008)	<8 years Adults	RT-PCR	3.1
Chan-it et al. [19]	Japan (urban 2008–2009)	Children of all age groups	RT-PCR	1.7
North America				
Mitchell et al. [79]	USA (urban, 1993–1994)	<3 years of age	EIA	20
Shastri et al. [99]	USA (urban, 1996–1997)	Children of all age groups. Mostly <3 years of age	EIA	6.9
Central America				
Cruz et al. [17]	Guatemala (rural, 1987–1989)	<3 years of age	EIA	38.6
Maldonado et al. [70]	Mexico (rural, 1992–1995)	<3 years of age	EIA and RT-PCR	61
South America				
Gaggero et al. [37]	Chile (urban, 1985–1987; 1993–1995)	<3 years of age	EIA	23.5
Espul et al. [27]	Argentina (urban, 1995–1998)	<3 years of age	EIA plus RT-PCR	3.7

(continued)

Table 1.1 (continued)

Study	Region (setting and period of study)	Age of study population	Detection method	Astrovirus prevalence (%)
Medina et al. [74]	Colombia (urban, 1997–1999)	<5 years of age	EIA	5.0
Gabbay et al. [36]	Brazil (urban, 1990–1992)	<2 years of age	EIA plus RT-PCR	5.8
Europe				
Bon et al. [10]	France (urban, 1995–1998)	<3 years of age	EIA	6.0
Putzker et al. [95]	Germany (urban, 2000)	All ages	EIA	1.2
De Grazia et al. [22]	Italy (urban, 1999–2000)	<2 years of age	EIA	3.1
Guix et al. [47]	Spain (urban, 1997–2000)	Children of all age groups. Mostly <5 years of age	RT-PCR	4.9
Tran et al. [107]	France (urban, 2007)	<5 years of age	EIA	1.8
Oceania				
Palombo et al. [91]	Australia (urban, 1995)	<5 years of age	Northern (RNA) Dot Blot	4.2

Colombia, and Venezuela have shown higher incidences in younger populations [10, 17, 43, 48, 74, 83]. However, in Guatemala and France [10, 17], the detection rate at the age of 2 was also high.

The incidence of the different classical HAstV serotypes has been determined in several parts of the world (Table 1.2). Worldwide, serotype 1 (HAstV-1) is the most prevalent type with few exceptions [43]. The second most common type occurrence depends on the geographic location [47, 83, 109]. Additionally, temporal intertypic variation cannot be ruled out due to the lack of data on the long-time evolution of HAstV types in a given location. It has been reported that children may suffer episodes of diarrhea caused by a serotype and a few months later become infected with another HAstV serotype, suggesting a lack of heterotypic immunity between the different antigenic types which may influence their prevalence [47, 52].

In one study showing the age distribution of the different HAstV types, more than 50% of the samples with HAstV-1 and HAstV-3 were from children under 2 years of age, while the majority of samples with HAstV-4 were from children older than 3 years [47]. This different age pattern suggests that the immune protection acquired against the most prevalent serotypes earlier in life fails to confer any protection against new serotypes or variants appearing thereafter.

There is no clear association between pathogenicity and classical HAstV serotype. However, HAstV-3 has been associated in one study [14] with more severe gastroenteritis and higher fecal virus shedding. In this same study, HAstV-3 is linked with persistent infections lasting up to 3 months [14]. Additionally, the involvement of the C-terminal nsP1a protein in viral RNA replication and thus in the RNA load in feces has been described [35, 44, 45], suggesting a correlation between hyper variable region (HVR) genotypes and pathogenicity. Interestingly, some HVR genotypes show higher viral numbers in stool and the aforementioned HAstV-3 associated with persistent gastroenteritis and higher virus shedding in feces [14] corresponds to one of these particular HVR genotypes (HVR genotype IV) [46]. It is also worth noting that HAstV-4 isolates may be distributed in two different HVR genotypes (V and VI), which differ in their level of excretion in stool [46], suggesting that the high-replication phenotype is related to the HVR genotype rather than to the serotype [119]. On the contrary, no differences in the clinical features of infection between genogroups A and B have been observed (unpublished results). Figure 1.1 shows the occurrence of classical HAstV genogroup A and genogroup B in pediatric samples (up to 15 years) from four hospitals in Barcelona, Spain. The average age was 29.6 and 18.6 months, and the median age 23.0 and 12.0 months, for patients with genogroup A and B infections, respectively.

Classical HAstV infections follow a seasonal pattern. In temperate regions, HAstV infections occur mostly in winter although cases may also be observed in summer [47, 62, 77, 83, 101, 110]. This seasonality may reflect the better stability of the virus in the cold period [1, 2]. The behavior of HAstV is somewhat similar to that of other viral gastroenteritis agents such as rotaviruses [23] and noroviruses [67], but different from that of adenovirus infections that occur throughout the year [92]. However, sometimes it is difficult to observe a distinct winter peak in each year [83]. A different distribution pattern is reported in tropical climates where

Table 1.2 Summary of the major studies on the prevalence of classical astrovirus serotypes

Study	Region (setting and period of study)	Detection method	1 (%)	2 (%)	3 (%)	4 (%)	5 (%)	6 (%)	7 (%)	8 (%)
North Africa										
Nafici et al. [86]	Egypt (rural, 1995–1998)	RT-PCR and sequencing	43.3	3.6	12.0	4.8	15.7	7.2	0	12.0
Asia										
Nguyen et al. [87]	Vietnam (urban, 2005–2006)	RT-PCR and sequencing	100	0	0	0	0	0	0	0
Verma et al. [112]	India (urban, 2004–2008)	RT-PCR and sequencing	67.7[a]	9.7[a]	0	0	6.5[a]	0	0	16
Chan-it et al. [19]	Japan (urban, 2008–2009)	RT-PCR and sequencing	91	0	9	0	0	0	0	0
North America										
Mitchell et al. [79]	USA (urban, 1993–1994)	RT-PCR and sequencing	55	17	[b]	[b]	[b]	[b]	[b]	[b]
South America										
Gabbay et al. [36]	Brazil (urban, 1990–1992)	RT-nested PCR	45.5	27.3	12.1	12.1	0	3	0	0
Espul et al. [27]	Argentina (urban, 1995–1998)	EIA and RT-PCR and sequencing	41	13	13	25	8	0	0	0
Europe										
Lee et al. [62]	United Kingdom (urban, 1976–1992)	IEM	64.9	11.3	9.3	11.3%	2.1	0.3	0.	0
Noel et al. [88]	United Kingdom (urban, 1981–1993)	EIA and IEM	66	0.6	6.2	4.3	0	[b]	[b]	0
Guix et al. [47]	Spain (urban, 1997–2000)	RT-PCR and sequencing	38	6	19	26	0	0	0	11
Oceania										
Palombo et al. [91]	Australia (urban, 1995)	RT-PCR and sequencing	85	0	0	15	0	0	0	0

[a]ORF1a/ORF2 Recombinant strains

[b]No data were available for the prevalence of these serotypes in this study

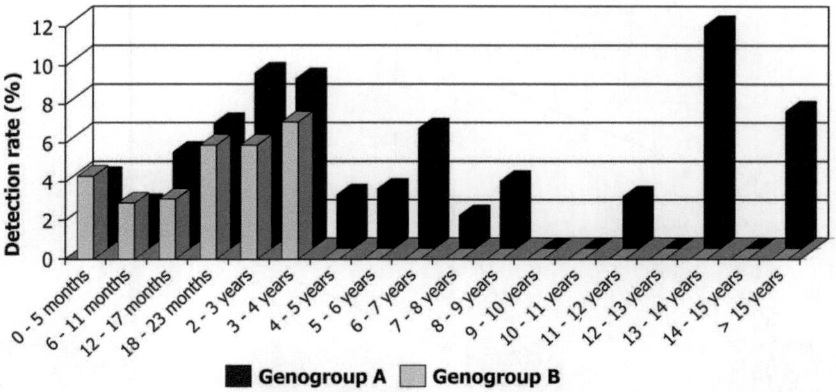

Fig. 1.1 HAstV in 2,291 fecal samples collected from infants and children with gastroenteritis at four hospitals from the Barcelona, Spain between May 1997 and April 2000. Primers A1 and A2 [47] and A1bis and A2 bis [82] were employed for HAstV, respectively. Positive samples were typed as described by Noel and coworkers [89]

HAstV detection rates tend to be higher in the rainy season [17, 70]. Additionally, some studies suggest a greater impact on the pediatric population in alternate years [47, 74, 83, 101].

Food and Environmental Transmission of Astroviruses

Human volunteer studies [58, 78] have shown that HAstV are transmitted through the fecal–oral route, although few adult recipient volunteers developed diarrheal illness with viral stool shedding, probably due to their age. It is well documented that food and water may act as vehicles for the transmission of human enteric viruses [11, 53]. Large outbreaks of astrovirus gastroenteritis of suspected foodborne origin have been reported in the literature [71, 80, 90]. Additionally, in institutions such as day care centers, hospitals, nurseries, schools, and military quarters, astroviruses may be transmitted through fecally contaminated environmental surfaces [2, 38]. Furthermore, infectious HAstV have been detected in water from an area where a concurrent gastroenteritis outbreak was reported [93], and a French epidemiology and risk assessment study suggests a role for waterborne astrovirus in the endemic level of digestive morbidity in the general population [40].

Figure 1.2 illustrates the potential routes of environmental transmission of HAstV. HAstV are shed in extremely high numbers, up to 10^{13} genome copies per gram, in the feces of infected individuals [14]. Since current treatment practices are unable to provide HAstV-free wastewater effluents, astroviruses are routinely introduced into the environment through the discharge of treated and untreated wastes [60, 82, 85], with one study describing a higher resistance to removal treatments for

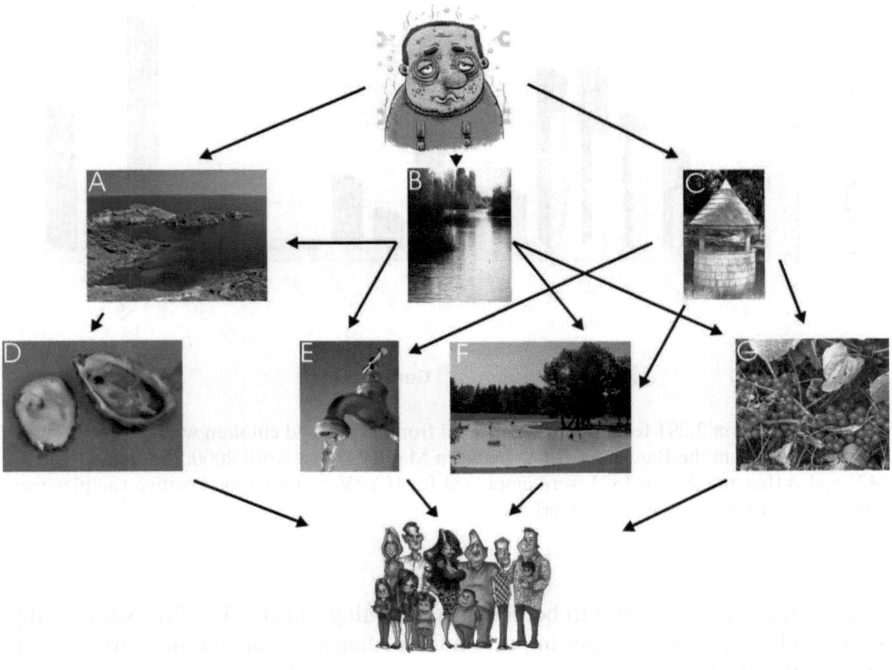

Fig. 1.2 Possible routes of environmental transmission of human astroviruses (see the text for details)

genogroup B than for genogroup A astrovirus [82]. The concurrent presence of identical HAstV strains in wastewater samples and in hospitalized patients has been evidenced in Pretoria [85]. Astroviruses may become contaminants of the marine environment [61] (Fig. 1.2a) or fresh water [93] (Fig. 1.2b). On the other hand, it is a fact that water reuse worldwide is increasing, and one major issue comprises reinjection of reclaimed wastewater into aquifers or ground water [5] (Fig. 1.2c).

A wide variety of foodstuffs may become contaminated by viruses during the farm-to-fork chain, either during the preharvest or postharvest stages. Among those foods at risk of preharvest contamination are bivalve mollusks grown in polluted waters (Fig. 1.2d), which have been reported to be capable of bearing HAstV [61] and have been implicated in gastroenteritis outbreaks [56]. Surface and ground waters are employed for public consumption (Fig. 1.2e) and poor water quality continues to pose a major threat to human health. The World Health Organization declared that 88% of the global burden of diarrheal disease is attributable to unsafe water supply, sanitation, and hygiene and is mostly concentrated on children in developing countries [116]. One study in Ghana reflects this situation and reports the detection by RT-PCR of HAstV RNA in tap water [25]. A significant amount of disease could be prevented especially in developing countries through better access to safe water supply, adequate sanitation facilities, and better hygiene practices.

Recreational activities in sewage-polluted waters have also been recognized as causes of infections. An outbreak of gastroenteritis occurred in Helsinki among children and adults after bathing in an outdoor wading pool (Fig. 1.2f) contaminated with noroviruses and astroviruses [72]. In addition, in a survey conducted in 1994 in the UK [84], 93% of surfers had HAstV antibodies compared with 22% of age-matched controls.

Salad crops, such as lettuce, green onions, and other greens, and soft fruits, such as raspberries and strawberries (Fig. 1.2g) are also susceptible to be contaminated at the preharvest stage through irrigation with sewage contaminated water and/or fertilization with sewage [94].

Postharvest contamination of food usually results from poor hygiene practices during food handling, and the foods most at risk are ready-to-eat food products including salads, sandwiches, and deli meats, which may be contaminated at the site of preparation by infected food handlers and are consumed uncooked or lightly cooked. For norovirus gastroenteritis, it has been shown that asymptomatic food workers are more often implicated than symptomatic workers [106]. There is evidence that HAstV infections can be asymptomatic not only in children but also in adults [75, 77] and it is particularly relevant to assess the importance of not only asymptomatic but also of presymptomatic and postsymptomatic shedders of astroviruses as sources of foodborne gastroenteritis.

Food handlers can also contaminate food when transferring viruses from contaminated surfaces to hands during preparation of ready-to-eat food or when transferring viruses from contaminated food to other ready-to-eat foods. Inanimate surfaces include contaminated utensils, e.g., chopping equipment, such as a dicer, cutting knives, and serving utensils, and may act as vehicles for HAstV transmission [2, 38]. The ability of HAstV to survive for extended time dried on inert surfaces supports this possibility [2]. As a matter of fact, one critical public health issue is whether once in the environment, infectious viruses are able to persist long enough and in high enough concentrations to pose a risk of infection. HAstV has been proven to be stable enough in drinking [1], fresh [13] and marine [13] water to be environmentally transmitted.

Chlorine is widely used for the routine inactivation of pathogenic microorganisms in water, particularly in drinking water. The behavior of astrovirus in the presence of free chlorine is not far from that of adenovirus, but different from that of rotavirus or hepatitis A virus, which show higher resistance to chlorine disinfection [1, 26]. In another study [102], the activity of halogeno-, cyano-, and amidino-isoflavenes, isoflavans, and flavans on the infectivity of human astroviruses was investigated. Although all drugs tested caused a dose-dependent reduction of the virus, the chloro derivatives were the most effective.

The efficacy of disinfection with alcohol on HAstV was investigated by Kurtz and coworkers in 1980 who recommended the use of preparations incorporating 90% ethanol [59]. Hands and fomites are among the principal vehicles for the transmission of infectious diseases, and antiseptics and disinfectants should be employed for the inactivation of pathogens in contaminated skin or surfaces, respectively, to reduce their transmission.

Table 1.3 Factors affecting foodborne and waterborne virus transmission factor

Virus stability
Amounts of virus shed/degree of viral contamination
Treatment of water/processing of food
Likelihood of infection after ingestion of one or a few virus particles
Susceptibility of the host

Table 1.3 lists the main factors affecting foodborne and waterborne virus spread which have been discussed earlier. In contrast with other pathogen or chemical contaminants, food and water are rarely tested for viral contamination. The main reason for this lack of virological analysis is the absence of legal regulations specifying such control, which is a direct consequence of several factors such as the important technical limitations that have hampered the development of reference methods for virus detection in water and food, and, last but not least, the high cost to be charged to a food product or drinking water. However, the benefits of reducing health problems through consumption of safe water and food products, which is particularly relevant in the present situation of global food trade, call for efforts in clearly defining which matrices should be tested for viruses, in the development of standardized reference methods targeted to specifically designed health-relevant viruses and in the formulation of guidelines to ensure the virological quality of selected commodities in specific scenarios.

Acknowledgments
These studies were funded in part by projects SGR: 2009SGR24 and the Catalonian Network of Biotechnology of the Generalitat de Catalunya and project Consolider: CDS2007-00016 of the Spanish Ministery of Science and Innovation.

References

1. Abad FX, Pinto RM, Villena C, Gajardo R, Bosch A. Astrovirus survival in drinking water. Appl Environ Microbiol. 1997;63:3119–22.
2. Abad FX, Villena C, Guix S, Caballero S, Pinto RM, Bosch A. Potential role of fomites in the vehicular transmission of human astroviruses. Appl Environ Microbiol. 2001;67:3904–7.
3. Aminu M, Ameh EA, Geyer A, Esona MD, Taylor MB, Steele AD. Role of astrovirus in intussusception in Nigerian infants. J Trop Pediatr. 2009;55:192–4.
4. Appleton H, Higgins PG. Letter: Viruses and gastroenteritis in infants. Lancet. 1975;1:1297.
5. Asano T, Cotruvo JA. Groundwater recharge with reclaimed municipal wastewater: health and regulatory considerations. Water Res. 2004;38:1941–51.
6. Basu G, Rossouw J, Sebunya TK, Gashe BA, de Beer M, Dewar JB, et al. Prevalence of rotavirus, adenovirus and astrovirus infection in young children with gastroenteritis in Gaborone, Botswana. East Afr Med J. 2003;80:652–5.
7. Belliot G, Laveran H, Monroe SS. Detection and genetic differentiation of human astroviruses: phylogenetic grouping varies by coding region. Arch Virol. 1997;142:1323–34.
8. Belliot G, Laveran H, Monroe SS. Outbreak of gastroenteritis in military recruits associated with serotype 3 astrovirus infection. J Med Virol. 1997;51:101–6.

9. Bjorkholm M, Celsing F, Runarsson G, Waldenstrom J. Successful intravenous immunoglobulin therapy for severe and persistent astrovirus gastroenteritis after fludarabine treatment in a patient with Waldenstrom's macroglobulinemia. Int J Hematol. 1995;62:117–20.
10. Bon F, Fascia P, Dauvergne M, Tenenbaum D, Planson H, Petion AM, et al. Prevalence of group A rotavirus, human calicivirus, astrovirus, and adenovirus type 40 and 41 infections among children with acute gastroenteritis in Dijon, France. J Clin Microbiol. 1999;37: 3055–8.
11. Bosch A. Human viruses in water. 1st ed. Amsterdam, Boston: Elsevier; 2007.
12. Bosch A, Méndez E, Krishna NK, Pantin-Jackwood M, Schultz-Cherry S, Monroe SS, et al. Astroviridae. In: King AMQ, Adams MJ, Carstens EB, Lefkowitz EJ, editors. Virus taxonomy. 9th ed. Oxford: Elsevier; 2011.
13. Bosch A, Pinto RM, Villena C, Abad FX. Persistence of human astrovirus in fresh and marine water. Water Sci Technol. 1997;35:243–7.
14. Caballero S, Guix S, El-Senousy WM, Calico I, Pinto RM, Bosch A. Persistent gastroenteritis in children infected with astrovirus: association with serotype-3 strains. J Med Virol. 2003;71:245–50.
15. Carter MJ, Willcocks MM. Astroviruses. 2006;eLS.doi:10.1038/npg.els.0004255.
16. Cox GJ, Matsui SM, Lo RS, Hinds M, Bowden RA, Hackman RC, et al. Etiology and outcome of diarrhea after marrow transplantation: a prospective study. Gastroenterology. 1994;107: 1398–407.
17. Cruz JR, Bartlett AV, Herrmann JE, Caceres P, Blacklow NR, Cano F. Astrovirus-associated diarrhea among guatemalan ambulatory rural children. J Clin Microbiol. 1992;30:1140–4.
18. Cubitt WD, Mitchell DK, Carter MJ, Willcocks MM, Holzel H. Application of electronmicroscopy, enzyme immunoassay, and RT-PCR to monitor an outbreak of astrovirus type 1 in a paediatric bone marrow transplant unit. J Med Virol. 1999;57:313–21.
19. Chan-it W, Thongprachum A, Okitsu S, Mizuguchi M, Ushijima H. Epidemiology and molecular characterization of sapovirus and astrovirus in Japan, 2008-2009. Jpn J Infect Dis. 2010; 63:302–3.
20. Chu DKW, Chin AWH, Smith GJ, Chan KH, Guan Y, Peiris JSM, et al. Detection of novel astroviruses in urban brown rats and previously known astroviruses in humans. J Gen Virol. 2010;91:2457–62.
21. De Benedictis P, Schultz-Cherry S, Burnham A, Cattoli G. Astrovirus infections in humans and animals - molecular biology, genetic diversity, and interspecies transmissions. Infect Genet Evol. 2011;11:1529–44.
22. De Grazia S, Giammanco GM, Colomba C, Cascio A, Arista S. Molecular epidemiology of astrovirus infection in Italian children with gastroenteritis. Clin Microbiol Infect. 2004; 10:1025–9.
23. Dey SK, Ushijima H, Phathammavong O, Chanit W, Okitsu S, Mizuguchi M, et al. Seasonal trend and serotype distribution of rotavirus infection in Japan, 1981-2008. Pediatr Infect Dis J. 2010;29:166–7.
24. Domingo E, Biebricher C, Eigen M, Holland J. Quasispecies and RNA virus evolution: principles and consequences. Austin, TX: Landes Bioscience; 2001.
25. Dongdem JT, Damanka S, Asmah R. Molecular isolation of human norovirus and astrovirus in tap water by RT- PCR. Int Res J Biochem Bioinform. 2011;1:131–8.
26. Espinosa AC, Mazari-Hiriart M, Espinosa R, Maruri-Avidal L, Mendez E, Arias CF. Infectivity and genome persistence of rotavirus and astrovirus in groundwater and surface water. Water Res. 2008;42:2618–28.
27. Espul C, Martinez N, Noel JS, Cuello H, Abrile C, Grucci S, et al. Prevalence and characterization of astroviruses in Argentinean children with acute gastroenteritis. J Med Virol. 2004; 72:75–82.
28. Fang ZY, Sun YP, Ye XH, Wang H, Zhang Q, Duan ZJ, et al. Astrovirus infection among hospitalized children with acute diarrhea in seven regions of China, 1998-2005. Zhonghua Liu Xing Bing Xue Za Zhi. 2006;27:673–6.

29. Finkbeiner SR, Allred AF, Tarr PI, Klein EJ, Kirkwood CD, Wang D. Metagenomic analysis of human diarrhea: viral detection and discovery. PLoS Pathog. 2008;4:e1000011.
30. Finkbeiner SR, Holtz LR, Jiang Y, Rajendran P, Franz CJ, Zhao G, et al. Human stool contains a previously unrecognized diversity of novel astroviruses. Virol J. 2009;6:161.
31. Finkbeiner SR, Kirkwood CD, Wang D. Complete genome sequence of a highly divergent astrovirus isolated from a child with acute diarrhea. Virol J. 2008;5:117.
32. Finkbeiner SR, Le BM, Holtz LR, Storch GA, Wang D. Detection of newly described astrovirus MLB1 in stool samples from children. Emerg Infect Dis. 2009;15:441–4.
33. Fodha I, Chouikha A, Peenze I, De Beer M, Dewar J, Geyer A, et al. Identification of viral agents causing diarrhea among children in the Eastern Center of Tunisia. J Med Virol. 2006; 78: 1198–203.
34. Foley B, O'Mahony J, Morgan SM, Hill C, Morgan JG. Detection of sporadic cases of Norwalk-like virus (NLV) and astrovirus infection in a single Irish hospital from 1996 to 1998. J Clin Virol. 2000;17:109–17.
35. Fuentes C, Guix S, Bosch A, Pinto RM. The C-terminal nsP1a protein of human astrovirus is a phosphoprotein that interacts with the viral polymerase. J Virol. 2011;85:4470–9.
36. Gabbay YB, Linhares AC, Cavalcante-Pepino EL, Nakamura LS, Oliveira DS, da Silva LD, et al. Prevalence of human astrovirus genotypes associated with acute gastroenteritis among children in Belem, Brazil. J Med Virol. 2007;79:530–8.
37. Gaggero A, O'Ryan M, Noel JS, Glass RI, Monroe SS, Mamani N, et al. Prevalence of astrovirus infection among Chilean children with acute gastroenteritis. J Clin Microbiol. 1998; 36:3691–3.
38. Gallimore CI, Taylor C, Gennery AR, Cant AJ, Galloway A, Lewis D, et al. Use of a heminested reverse transcriptase PCR Assay for detection of astrovirus in environmental swabs from an outbreak of gastroenteritis in a pediatric primary immunodeficiency unit. J Clin Microbiol. 2005;43:3890–4.
39. Glass RI, Noel J, Mitchell D, Herrmann JE, Blacklow NR, Pickering LK, et al. The changing epidemiology of astrovirus-associated gastroenteritis: a review. Arch Virol Suppl. 1996; 12:287–300.
40. Gofti-Laroche L, Gratacap-Cavallier B, Demanse D, Genoulaz O, Seigneurin JM, Zmirou D. Are waterborne astrovirus implicated in acute digestive morbidity (E.MI.R.A. study)? J Clin Virol. 2003;27:74–82.
41. Gray JJ, Wreghitt TG, Cubitt WD, Elliot PR. An outbreak of gastroenteritis in a home for the elderly associated with astrovirus type 1 and human calicivirus. J Med Virol. 1987; 23:377–81.
42. Grohmann GS, Glass RI, Pereira HG, Monroe SS, Hightower AW, Weber R, et al. Enteric viruses and diarrhea in HIV-infected patients. Enteric opportunistic infections working group. N Engl J Med. 1993;329:14–20.
43. Guerrero ML, Noel JS, Mitchell DK, Calva JJ, Morrow AL, Martinez J, et al. A prospective study of astrovirus diarrhea of infancy in Mexico city. Pediatr Infect Dis J. 1998;17:723–7.
44. Guix S, Caballero S, Bosch A, Pinto RM. C-terminal nsP1a protein of human astrovirus colocalizes with the endoplasmic reticulum and viral RNA. J Virol. 2004;78:13627–36.
45. Guix S, Caballero S, Bosch A, Pinto RM. Human astrovirus C-terminal nsP1a protein is involved in RNA replication. Virology. 2005;333:124–31.
46. Guix S, Caballero S, Fuentes C, Bosch A, Pinto RM. Genetic analysis of the hypervariable region of the human astrovirus nsP1a coding region: design of a new RFLP typing method. J Med Virol. 2008;80:306–15.
47. Guix S, Caballero S, Villena C, Bartolome R, Latorre C, Rabella N, et al. Molecular epidemiology of astrovirus infection in Barcelona, Spain. J Clin Microbiol. 2002;40:133–9.
48. Herrmann JE, Taylor DN, Echeverria P, Blacklow NR. Astroviruses as a cause of gastroenteritis in children. N Engl J Med. 1991;324(25):1757–60.
49. Jakab F, Walter JE, Berke T, Matson DO, Mitchell DK, Szucs G. Molecular characterization and sequence analysis of human astroviruses circulating in Hungary. FEMS Immunol Med Microbiol. 2003;39:97–102.

50. Kapoor A, Li L, Victoria J, Oderinde B, Mason C, Pandey P, et al. Multiple novel astrovirus species in human stool. J Gen Virol. 2009;90:2965–72.
51. Kiulia NM, Mwenda JM, Nyachieo A, Nyaundi JK, Steele AD, Taylor MB. Astrovirus infection in young Kenyan children with diarrhoea. J Trop Pediatr. 2007;53:206–9.
52. Koopmans MP, Bijen MH, Monroe SS, Vinje J. Age-stratified seroprevalence of neutralizing antibodies to astrovirus types 1 to 7 in humans in The Netherlands. Clin Diagn Lab Immunol. 1998;5:33–7.
53. Koopmans MPG, Cliver DO, Bosch A. Food-borne viruses: progress and challenges. Washington, DC: ASM Press; 2008.
54. Kriston S, Willcocks MM, Carter MJ, Cubitt WD. Seroprevalence of astrovirus types 1 and 6 in London, determined using recombinant virus antigen. Epidemiol Infect. 1996;117: 159–64.
55. Kurtz J, Lee T. Astrovirus gastroenteritis age distribution of antibody. Med Microbiol Immunol. 1978;166:227–30.
56. Kurtz JB, Lee TW. Astroviruses: human and animal. Ciba Found Symp. 1987;128:92–107.
57. Kurtz JB, Lee TW. Human astrovirus serotypes. Lancet. 1984;2:1405.
58. Kurtz JB, Lee TW, Craig JW, Reed SE. Astrovirus infection in volunteers. J Med Virol. 1979;3:221–30.
59. Kurtz JB, Lee TW, Parsons AJ. The action of alcohols on rotavirus, astrovirus and enterovirus. J Hosp Infect. 1980;1:321–5.
60. Le Cann P, Ranarijaona S, Monpoeho S, Le Guyader F, Ferre V. Quantification of human astroviruses in sewage using real-time RT-PCR. Res Microbiol. 2004;155:11–5.
61. Le Guyader F, Haugarreau L, Miossec L, Dubois E, Pommepuy M. Three-year study to assess human enteric viruses in shellfish. Appl Environ Microbiol. 2000;66:3241–8.
62. Lee TW, Kurtz JB. Prevalence of human astrovirus serotypes in the Oxford region 1976-92, with evidence for two new serotypes. Epidemiol Infect. 1994;112:187–93.
63. Lee TW, Kurtz JB. Serial propagation of astrovirus in tissue culture with the aid of trypsin. J Gen Virol. 1981;57:421–4.
64. Lewis DC, Lightfoot NF, Cubitt WD, Wilson SA. Outbreaks of astrovirus type 1 and rotavirus gastroenteritis in a geriatric in-patient population. J Hosp Infect. 1989;14:9–14.
65. Liste MB, Natera I, Suarez JA, Pujol FH, Liprandi F, Ludert JE. Enteric virus infections and diarrhea in healthy and human immunodeficiency virus-infected children. J Clin Microbiol. 2000;38(8):2873–7.
66. Liu M-Q, Peng J-S, Tang L, Zhou Y, Yang B-F, Wang Y-H, et al. Identification of new subtype of astrovirus type 3 from an infant with diarrhea in Wuhan, China. Virology. 2008; 375: 301–6.
67. Lopman B, Zambon M, Brown DW. The evolution of norovirus, the "gastric flu". PLoS Med. 2008;5:e42.
68. Lukashov VV, Goudsmit J. Evolutionary relationships among Astroviridae. J Gen Virol. 2002;83:1397–405.
69. Madeley CR, Cosgrove BP. 28 nm particles in faeces in infantile gastroenteritis. Lancet. 1975;6:451–2.
70. Maldonado Y, Cantwell M, Old M, Hill D, Sanchez ML, Logan L, et al. Population-based prevalence of symptomatic and asymptomatic astrovirus infection in rural Mayan infants. J Infect Dis. 1998;178:334–9.
71. Matsui SM, Greenberg HB. Medical management of foodborne viral gastroenteriti. In: Hui YH, Gorham JR, Murrell KD, Cliver DOE, editors. Diseases caused by viruses, parasites, and fungi, Foodborne disease handbook, vol. 2. New York: Dekker; 1994.
72. Maunula L, Kalso S, von Bonsdorff CH, Ponka A. Wading pool water contaminated with both noroviruses and astroviruses as the source of a gastroenteritis outbreak. Epidemiol Infect. 2004;132:737–43.
73. Mead PS, Slutsker L, Dietz V, McCaig LF, Bresee JS, Shapiro C, et al. Food-related illness and death in the United States. Emerg Infect Dis. 1999;5:607–25.

74. Medina SM, Gutierrez MF, Liprandi F, Ludert JE. Identification and type distribution of astroviruses among children with gastroenteritis in Colombia and Venezuela. J Clin Microbiol. 2000;38(9):3481–3.
75. Mendez-Toss M, Griffin DD, Calva J, Contreras JF, Puerto FI, Mota F, et al. Prevalence and genetic diversity of human astroviruses in Mexican children with symptomatic and asymptomatic infections. J Clin Microbiol. 2004;42:151–7.
76. Mendez-Toss M, Romero-Guido P, Munguia ME, Mendez E, Arias CF. Molecular analysis of a serotype 8 human astrovirus genome. J Gen Virol. 2000;81:2891–7.
77. Mendez E, Arias CF. Astroviruses. In: Knipe DM, Howley PM, editors. Fields virology, vol. 5. Philadelphia: Lippincott Williams and Wilkins; 2007.
78. Midthun K, Greenberg HB, Kurtz JB, Gary GW, Lin FY, Kapikian AZ. Characterization and seroepidemiology of a type 5 astrovirus associated with an outbreak of gastroenteritis in Marin County, California. J Clin Microbiol. 1993;31:955–62.
79. Mitchell DK, Matson DO, Jiang X, Berke T, Monroe SS, Carter MJ, et al. Molecular epidemiology of childhood astrovirus infection in child care centers. J Infect Dis. 1999;180:514–7.
80. Mitchell DK, Monroe SS, Jiang X, Matson DO, Glass RI, Pickering LK. Virologic features of an astrovirus diarrhea outbreak in a day care center revealed by reverse transcriptase-polymerase chain reaction. J Infect Dis. 1995;172:1437–44.
81. Monroe SS, Jiang B, Stine SE, Koopmans M, Glass RI. Subgenomic RNA sequence of human astrovirus supports classification of Astroviridae as a new family of RNA viruses. J Virol. 1993;67:3611–4.
82. Morsy El-Senousy W, Guix S, Abid I, Pinto RM, Bosch A. Removal of astrovirus from water and sewage treatment plants, evaluated by a competitive reverse transcription-PCR. Appl Environ Microbiol. 2007;73:164–7.
83. Mustafa H, Palombo EA, Bishop RF. Epidemiology of astrovirus infection in young children hospitalized with acute gastroenteritis in Melbourne, Australia, over a period of four consecutive years, 1995 to 1998. J Clin Microbiol. 2000;38(3):1058–62.
84. Myint S, Manley R, Cubitt D. Viruses in bathing waters. Lancet. 1994;343:1640–1.
85. Nadan S, Walter JE, Grabow WO, Mitchell DK, Taylor MB. Molecular characterization of astroviruses by reverse transcriptase PCR and sequence analysis: comparison of clinical and environmental isolates from South Africa. Appl Environ Microbiol. 2003;69:747–53.
86. Naficy AB, Rao MR, Holmes JL, Abu-Elyazeed R, Savarino SJ, Wierzba TF, et al. Astrovirus diarrhea in Egyptian children. J Infect Dis. 2000;182:685–90.
87. Nguyen TA, Hoang L, Phamle D, Hoang KT, Mizuguchi M, Okitsu S, et al. Identification of human astrovirus infections among children with acute gastroenteritis in the southern part of Vietnam during 2005-2006. J Med Virol. 2008;80:298–305.
88. Noel J, Cubitt D. identification of astrovirus serotypes from children treated at the hospitals for sick children, London 1981-93. Epidemiol Infect. 1994;113:153–9.
89. Noel JS, Lee TW, Kurtz JB, Glass RI, Monroe SS. Typing of human astroviruses from clinical isolates by enzyme immunoassay and nucleotide sequencing. J Clin Microbiol. 1995;33: 797–801.
90. Oishi I, Yamazaki K, Kimoto T, Minekawa Y, Utagawa E, Yamazaki S, et al. A large outbreak of acute gastroenteritis associated with astrovirus among students and teachers in Osaka, Japan. J Infect Dis. 1994;170:439–43.
91. Palombo EA, Bishop RF. Annual incidence, serotype distribution, and genetic diversity of human astrovirus isolates from hospitalized children in Melbourne, Australia. J Clin Microbiol. 1996;34:1750–3.
92. Parashar, U. D., and R. I. Glass (2003) Viral causes of gastroenteritis. In: J. G. Ulrich Desselberger (ed.), Perspectives in Medical Virology, Elsevier
93. Pintó RM, Abad FX, Gajardo R, Bosch A. Detection of infectious astroviruses in water. Appl Environ Microbiol. 1996;62:1811–3.
94. Pintó RM, Bosch A. Rethinking virus detection in food. In: Koopmans M, Cliver DO, Bosch A, editors. Foodborne viruses: progress and challenges. Washington, DC: ASM Press; 2008.

95. Putzker M, Sauer H, Kirchner G, Keksel O, Malic A. Community acquired diarrhea–the incidence of astrovirus infections in Germany. Clin Lab. 2000;46:269–73.
96. Quan PL, Wagner TA, Briese T, Torgerson TR, Hornig M, Tashmukhamedova A, et al. Astrovirus encephalitis in boy with X-linked agammaglobulinemia. Emerg Infect Dis. 2010; 16:918–25.
97. Reither K, Ignatius R, Weitzel T, Seidu-Korkor A, Anyidoho L, Saad E, et al. Acute childhood diarrhoea in northern Ghana: epidemiological, clinical and microbiological characteristics. BMC Infect Dis. 2007;7:104.
98. Rivera R, Nollens HH, Venn-Watson S, Gulland FMD, Wellehan Jr JFX. Characterization of phylogenetically diverse astroviruses of marine mammals. J Gen Virol. 2010;91:166–73.
99. Shastri S, Doane AM, Gonzales J, Upadhyayula U, Bass DM. Prevalence of astroviruses in a children's hospital. J Clin Microbiol. 1998;36:2571–4.
100. Silva PA, Cardoso DD, Schreier E. Molecular characterization of human astroviruses isolated in Brazil, including the complete sequences of astrovirus genotypes 4 and 5. Arch Virol. 2006;151:1405–17.
101. Steele AD, Basetse HR, Blacklow NR, Herrmann JE. Astrovirus infection in South Africa: a pilot study. Ann Trop Paediatr. 1998;18:315–9.
102. Superti F, Seganti L, Orsi N, Desideri N, Stein ML, Tinari A, et al. In vitro effect of synthetic flavanoids on astrovirus infection. Antiviral Res. 1990;13:201–8.
103. Svenungsson B, Lagergren A, Ekwall E, Evengard B, Hedlund KO, Karnell A, et al. Enteropathogens in adult patients with diarrhea and healthy control subjects: a 1-year prospective study in a Swedish clinic for infectious diseases. Clin Infect Dis. 2000;30:770–8.
104. Taylor MB, Marx FE, Grabow WOK. Rotavirus, astrovirus and adenovirus associated with an outbreak of gastroenteritis in a South African child care centre. Epidemiol Infect. 1997; 119:227–30.
105. Taylor MB, Walter J, Berke T, Cubitt WD, Mitchell DK, Matson DO. Characterisation of a South African human astrovirus as type 8 by antigenic and genetic analyses. J Med Virol. 2001;64:256–61.
106. Todd EC, Greig JD, Bartleson CA, Michaels BS. Outbreaks where food workers have been implicated in the spread of foodborne disease. Part 2. Description of outbreaks by size, severity, and settings. J Food Prot. 2007;70:1975–93.
107. Tran A, Talmud D, Lejeune B, Jovenin N, Renois F, Payan C, et al. Prevalence of rotavirus, adenovirus, norovirus, and astrovirus infections and coinfections among hospitalized children in northern France. J Clin Microbiol. 2010;48:1943–6.
108. Ulloa JC, Gutierrez MF. Genomic analysis of two ORF2 segments of new porcine astrovirus isolates and their close relationship with human astroviruses. Can J Microbiol. 2010; 56:569–77.
109. Unicomb LE, Banu NN, Azim T, Islam A, Bardhan PK, Faruque AS, et al. Astrovirus infection in association with acute, persistent and nosocomial diarrhea in Bangladesh. Pediatr Infect Dis J. 1998;17:611–4.
110. Utagawa ET, Nishizawa S, Sekine S, Hayashi Y, Ishihara Y, Oishi I, et al. Astrovirus as a cause of gastroenteritis in Japan. J Clin Microbiol. 1994;32:1841–5.
111. van Hemert FJ, Berkhout B, Lukashov VV. Host-related nucleotide composition and codon usage as driving forces in the recent evolution of the Astroviridae. Virology. 2007;361: 447–54.
112. Verma H, Chitambar SD, Gopalkrishna V. Astrovirus associated acute gastroenteritis in western India: predominance of dual serotype strains. Infect Genet Evol. 2010;10:575–9.
113. Walter JE, Briggs J, Guerrero ML, Matson DO, Pickering LK, Ruiz-Palacios G, et al. Molecular characterization of a novel recombinant strain of human astrovirus associated with gastroenteritis in children. Arch Virol. 2001;146:2357–67.
114. Walter JE, Mitchell DK. Astrovirus infection in children. Curr Opin Infect Dis. 2003;16: 247–53.
115. Walter JE, Mitchell DK. Role of astroviruses in childhood diarrhea. Curr Opin Pediatr. 2000; 12:275–9.
116. WHO. Guidelines for Drinking-water Quality, Third Edition, Volume 1: Recommendations, p. 515 pp. Geneva, Switzerland:World Health Organization; 2004.

117. Wolfaardt M, Kiulia NM, Mwenda JM, Taylor MB. Evidence of a recombinant wild-type human astrovirus strain from a Kenyan child with gastroenteritis. J Clin Microbiol. 2011; 49:728–31.
118. Wood DJ, David TJ, Chrystie IL, Totterdell B. Chronic enteric virus infection in two T-cell immunodeficient children. J Med Virol. 1988;24:435–44.
119. Zhang Z, Mitchell DK, Afflerbach C, Jakab F, Walter J, Zhang YJ, et al. Quantitation of human astrovirus by real-time reverse-transcription-polymerase chain reaction to examine correlation with clinical illness. J Virol Methods. 2006;134:190–6.

Chapter 2
Replication Cycle of Astroviruses

Ernesto Méndez, Andrea Murillo, Rodrigo Velázquez, Andrew Burnham, and Carlos F. Arias

Abstract Astrovirus infections cause gastroenteritis in mammals and have been identified as causative agents of diverse pathologies in birds such as hepatitis in ducks and poult enteritis mortality syndrome (PEMS), which causes enteritis and thymic and bursal atrophy in turkeys. Human astroviruses are recognized as the second leading cause of childhood viral gastroenteritis worldwide. Eight traditional astrovirus serotypes have been identified in humans, but recently novel astrovirus strains isolated from humans have been associated with diseases other than gastroenteritis. Herein we summarize our current knowledge of the astrovirus life cycle. Though there are gaps in our understanding of astrovirus replication, similarities can be drawn from *Picornaviridae* and *Caliciviridae* virus families. There are, however, unique characteristics of the astrovirus life cycle, including intracellular proteolytic processing of viral particles by cellular caspases, which has been shown to be required for the maturation and exit of viral progeny.

Introduction

The family *Astroviridae* includes human and animal viruses that share a characteristic icosahedral morphology; they are nonenveloped and their genome is composed of positive-sense, single-stranded RNA, whose organization distinguishes them

E. Méndez, Ph.D. • A. Murillo, M.Sc. • R. Velázquez, M.S. • C.F. Arias, Ph.D. (✉)
Genética del Desarrollo y Fisiología Molecular, Instituto de Biotecnología,
Universidad Nacional Autónoma de México, Av. Universidad # 2001,
Colonia Chamilpa, Cuernavaca 62210, Mexico
e-mail: murilloa@ibt.unam.mx; rovm85@hotmail.com; arias@ibt.unam.mx

A. Burnham, Ph.D.
Department of Infectious Diseases, St. Jude Children's Research Hospital,
262 Danny Thomas Place, Memphis, TN 38105, USA
e-mail: andrew.burnham@stjude.org

S. Schultz-Cherry (ed.), *Astrovirus Research: Essential Ideas, Everyday Impacts, Future Directions*, DOI 10.1007/978-1-4614-4735-1_2,

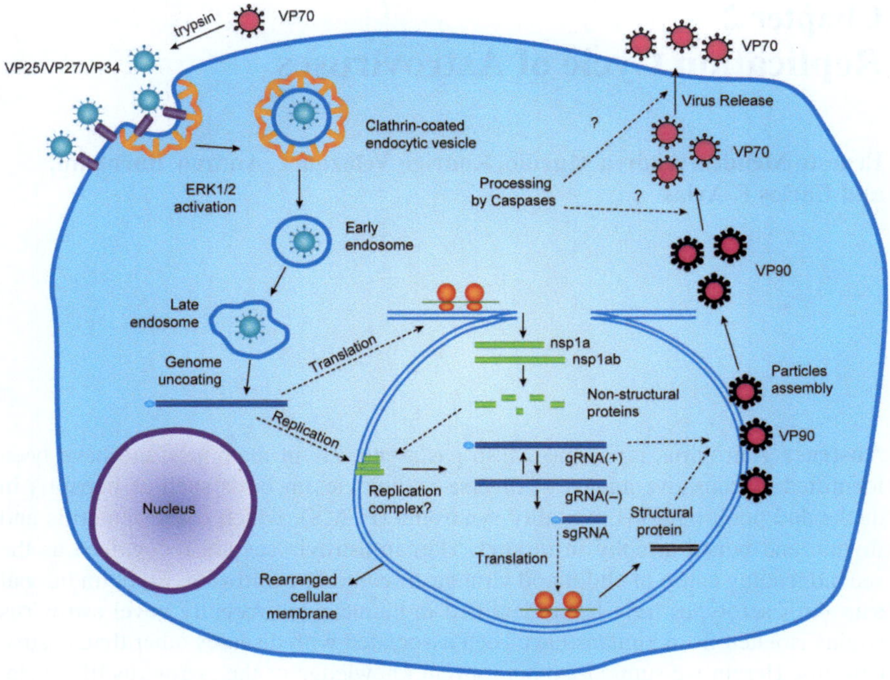

Fig. 2.1 Replication cycle of astroviruses. See details in the text

from other virus families with similar properties. Efficient propagation of astrovirus strains from both human and animal origin in cultured cells has permitted advances in the characterization of their life cycle (Fig. 2.1). After astrovirus cell entry, the genomic RNA is uncoated and is then translated into a polyprotein precursor that is subsequently cleaved into the proteins required for replication of the virus genome and assembly of progeny viruses. These processes most likely take place in association with cellular membranes, and completion of the virus life cycle requires the action of caspases for the virus particles to exit the cell.

Genome Structure and Organization

Astroviruses have a positive-sense, single-stranded RNA genome that varies in length from 6.17 kb for human astrovirus (HAstV) MLB-1 to 7.72 kb for duck astrovirus-2, excluding the poly(A) tail at the 3′-end [54]. The viral genome contains 5′ and 3′ untranslated regions (UTR) and three open reading frames (ORFs) (Fig. 2.2). The two ORFs located toward the 5′-end of the genome, designated ORF1a and ORF1b, encode nonstructural proteins that are presumed to be involved

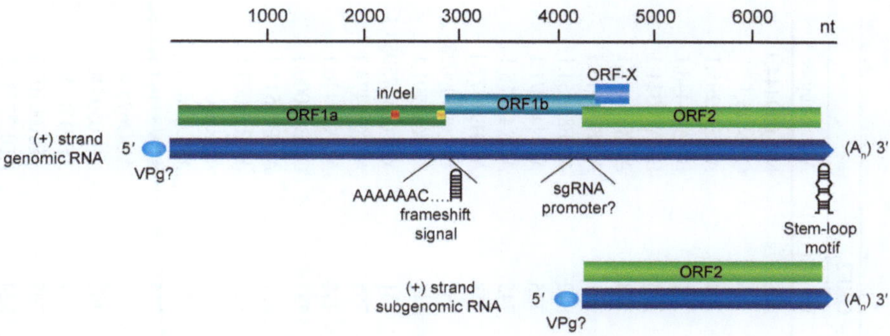

Fig. 2.2 Genome organization of human astrovirus. The genomic RNA, of approximately 6.8 kb, contains three recognized open reading frames (ORF1a, ORF1b, and ORF2). ORF-X has been proposed as functional, given its conservation among astrovirus The genome contains three elements conserved among all members of the family: the frameshift signal (*blue square*), the sequence upstream ORF2 that putatively acts as promoter for synthesis of the subgenomic (sg) RNA, and the stem-loop at the 3′-end. Also shown are the insertion/deletion (In/del) region (*red square*) and the presence of a poly(A) tail at the 3′-end. A putative VPg protein is depicted as bound to the 5′-end of the genomic and subgenomic RNAs

in RNA transcription and replication based on the sequence motifs they contain. The third ORF, found at the 3′-end of the genome, designated ORF2, is common to both the genomic and subgenomic RNAs and encodes the capsid proteins [50, 76]. An additional ORF (ORF-X) of 91–122 codons, contained within ORF2 in a +1 reading frame, has been described in all HAstV and some other mammalian viruses (Fig. 2.2) [18]. Its initiation codon is located 41–50 nucleotides downstream the ORF2 AUG and might be translated through a leaky scanning mechanism. It remains to be determined whether the putative protein encoded in ORF-X is synthesized in HAstV-infected cells and its significance for virus replication.

The length of all ORFs varies among different astrovirus strains, with the largest variation observed in ORF1a. The variation in this ORF is mainly due to insertions or deletions (in/del regions) present near the 3′-end [30] that have been associated with the cell line used to adapt the virus for growth in cell culture (Table 2.1) [73]. ORF1b is similar in length among all astroviruses. Comparative analysis of ORF2 among astrovirus has been more complete, because the sequences of a large number of strains of diverse origin have been determined. Its size is not as variable as that of ORF1a (Table 2.1), but it shows the highest sequence variability (Tables 2.2, 2.3, and 2.4).

A comparison of the complete genome sequences of astrovirus strains reported thus far has recognized common features. An overlap of 10–148 nt in the genome of mammalian viruses has been found between ORF1a and ORF1b. These two ORFs also overlap in avian viruses, but this overlapping is only 12–45 nt long [54]. The overlapping region contains signals that are essential for translation of the viral RNA polymerase encoded in ORF1b through a frameshift mechanism. During infection of susceptible cells, two positive-sense RNA species have been observed: the full-length genomic RNA (gRNA), and a subgenomic RNA (sgRNA, of ~2.4 kb) [38, 49].

Table 2.1 Length variation of astrovirus polyproteins

		nsp1a		nsp1b		ORF2	
Animal origin	Virus strain	Length (aa)	Acc. number	Length (aa)	Acc. number	Length (aa)	Acc. number
Human	HAstrovirus-1	920	AAC34715.1	519	AAC34716.1	786	NP_059444.1
	HAstrovirus-2					796	Q82446.1
	HAstrovirus-3	928	ADE09293.1	515	AEN74891.1	794	ADE09295.1
	HAstrovirus-4	920	AAY84777.1	515	AAW51878.1	771	BAA93448.1
	HAstrovirus-5	950	Q4TWH9.1	515	AAY46273.1	783	BAA90311.1
	HAstrovirus-6	921	ACX83562.2	515	ACV92106.1	778	ACX83561.1
	HAstrovirus-7					791	AAK31913.1
	HAstrovirus-8	921	Q9IFX3.1	518	AAF85963.1	842	AAF85964.1
	HAstrovirusMLB1	787	ACN44169.1	511	YP_002290967.1	755	ACN44171.1
	HAstrovirusMLB2	787	AER41412.1	511	ACX69837.1	745	AER41415.1
	HAstrovirus-SG	885	ADH93576.1	508	ADH93575.1	758	ADH93577.1
	HAstrovirus-VA1	886	YP_003090287.1	487	ACR23347.1	758	YP_003090288.1
	HAstrovirus-VA2	845	ACX83590.1	528	ACX69838.1	642	ACX83591.1
	HMOAstrovirus-A	874	YP_003275952.1	487	YP_003275951.1	731	YP_003275953.1
	HMOAstrovirus-B					755	ACX85474.1
	HMOAstrovirus-C					758	ACX85476.1
Porcine	PAstrovirus					783	CAB95000.1
	PAstrovirus-2	824	AER29998.1	498	AER29999.1	851	ADP21511.1
	PAstrovirus-4	850	AER30007.1	492	AER30008.1	825	AER30009.1
	PAstrovirus-5	872	AER30001.1	491	AER30002.1	735	AER30003.1
Bovine	BAstrovirus	815	AED89607.1	452	AED89583.1	772	AED89600.1
Deer	CcAstrovirus-1					755	AAV37187.1
	CcAstrovirus-2					749	ADO67581.1
Ovine	OAstrovirus	844	NP_059944.1	523	CAB95003.1	762	NP_059946.1
	OAstrovirus-2					737	AEP17838.1
Canine	CaAstrovirus					765	AEX00102.1
	SHDAstrovirus					824	ADL27736.1

Feline	FAstrovirus					816	AAC13556.1
Bat	BatAstrovirus-1					850	ACF75865.1
Mouse	MouseAstrovirus	849	YP_004782205.1			707	YP_004782206.1
Rat	RatAstrovirus			516	ADJ38390.1	810	ADJ38394.1
Mink	MAstrovirus	874	NP_795334.1	507	ADR65075.1	775	ADR65076.1
Sea lion	CslAstrovirus-1					780	ACR54272.1
	CslAstrovirus-2					859	ACR54274.1
	CslAstrovirus-4	912	AEM37614.1	530	AEM37615.1	828	AEM37616.1
	CslAstrovirus-5	929	AEM37617.1	515	AEM37618.1	766	AEM37619.1
	CslAstrovirus-6			530	AEM37621.1	840	AEM37622.1
	CslAstrovirus-7	888	AEM37623.1	535	AEM37624.1	839	AEM37625.1
	CslAstrovirus-8			530	AEM37627.1	825	AEM37628.1
	CslAstrovirus-9	888	AEM37629.1	535	AEM37630.1	838	AEM37631.1
	CslAstrovirus-10	888	AEM37632.1	536	AEM37633.1	825	AEM37634.1
	CslAstrovirus-11	934	AEM37638.1	515	AEM37639.1	752	AEM37640.1
Dolphin	BdAstrovirus-1					826	ACR54280.1
	BdAstrovirus-3					730	ADX97506.1
Whale	MWAstrovirus-2					779	ADX97522.1
Pigeon	FPAstrovirus					673	CBY02488.1
	WPAstrovirus					676	CBY02492.1
Duck	DAstrovirus	1,240	ADB79808.1	516	ADB79809.1	731	YP_002728003.1
Chicken	ANV-1	1,005	P0C6K7.1	507	Q9JGF2.1	680	ADG45753.1
	ANV-2					680	AEB15599.1
	Castrovirus	1,139	AEE88303.1	516	AEE88304.1	743	AEE88305.1
Turkey	Tastrovirus	1,126	ABX46576.1	517	ABX46574.1	723	ABX46584.1
	TAstrovirus-2	1,125	ABX46564.1	527	AAF60952.1	725	ABX46566.1
	TAstrovirus-3					724	AAV37187.1

Table 2.2 Sequence diversity of protein nsp1a of astroviruses[a]

	Virus	H-1	H-3	H-4	H-5	H-6	H-8	H-MLB1	H-MLB2	H-SG	H-VA1	H-VA2	HMO-A	P-2	P-4	P-5	B	O	Mouse	M	Csl-4	Csl-5	Csl-7	Csl-9	Csl-10	Csl-11	D	ANV-1	C	T	T-2
Human	HAstV-1	100																													
	HAstV-3	95	100																												
	HAstV-4	97	95	100																											
	HAstV-5	95	87	96	100																										
	HAstV-6	88	87	88	87	100																									
	HAstV-8	96	97	97	96	88	100																								
	HAstVMLB1	30	37	34	34	32	37	100																							
	HAstVMLB2	31	32	32	32	32	32	82	100																						
	HAstV-SG	26	26	26	26	26	27	28	29	100																					
	HAstV-VA1	29	28	27	27	28	28	30	29	97	100																				
	HAstV-VA2	26	27	26	27	26	27	30	31	66	66	100																			
	HMOAstV-A	27	28	27	26	27	27	29	30	65	65	98	100																		
Porcine	PAstV-2	32	33	33	33	33	33	30	30	31	28	30	30	100																	
	PAstV-4	32	32	32	32	33	32	34	31	30	30	28	31	36	100																
	PAstV-5	25	26	26	26	26	26	25	26	26	26	24	25	29	25	100															
Bovine	BAstV	29	29	29	29	29	29	30	29	29	29	32	32	71	34	30	100														
Ovine	OAstV	31	42	28	31	27	29	28	29	40	41	44	45	28	29	25	33	100													
Mouse	MouseAstV	26	27	27	26	26	26	28	27	25	26	26	26	27	26	25	27	28	100												
Mink	MAstV	29	29	28	29	29	28	29	29	41	41	42	43	29	27	28	30	40	26	100											
Sea lion	CslAstV-4	43	44	44	43	45	45	34	30	28	29	29	30	35	33	25	29	27	26	27	100										
	CslAstV-5	75	75	76	75	76	76	35	34	29	30	30	31	28	34	25	29	31	27	28	45	100									
	CslAstV-7	47	47	47	47	46	48	31	30	27	27	30	30	28	34	27	28	29	27	29	71	45	100								
	CslAstV-9	47	47	47	47	46	48	33	30	28	29	30	30	29	34	28	28	29	28	28	70	45	99	100							
	CslAstV-10	47	47	47	47	46	48	33	32	27	29	31	31	28	34	28	28	28	27	28	70	45	98	98	100						
	CslAstV-11	74	74	75	74	74	74	34	34	29	29	31	31	29	33	27	29	34	25	29	48	86	54	54	49	100					
Duck	DAstV	25	25	26	25	27	26	28	24	25	24	25	21	49	25	33	25	21	23	26	27	24	28	28	27	28	100				
Chicken	ANV-1	26	28	27	29	29	28	29	27	27	25	27	27	31	23	29	27	23	25	28	29	28	27	27	27	27	42	100			
	CAstV	25	26	26	25	23	25	25	24	25	25	23	24	27	26	25	25	24	24	25	24	26	24	24	23	25	59	28	100		
Turkey	TAstV	25	24	25	25	25	25	25	26	27	27	27	27	26	24	23	25	24	25	26	25	25	24	24	24	24	43	29	45	100	
	TAstV-2	25	24	25	24	27	24	25	26	28	28	28	28	27	23	26	26	24	24	27	25	24	24	24	22	23	50	29	45	98	100

[a]All available nsp1a sequences at the NCBI database were downloaded for each astrovirus strain. The sequences were pairwised compared, and their percent amino acid identities are shown in the figure.

Both RNAs are transcribed from the negative-sense full-length complement of the genomic RNA. Based on the transcription initiation site determined for the sgRNA in HAstV-1 and HAstV-2 [50, 76], ORF1b and ORF2 overlap by 8 nt; however, the length of this overlapping may vary, and it is not present in a duck astrovirus, in which it is 24 nt apart [19]. The synthesis of sgRNA requires an internal subgenomic promoter in the negative-sense RNA, and a highly conserved sequence in mammalian astroviruses around the ORF2 start codon that has been suggested to be part of the promoter for synthesis of the sgRNA [36]. This sequence is also conserved in avian astroviruses, but is slightly different in sequence and length [54]. This conserved sequence has a partial identity with the 5′-end of the gRNA [36], suggesting it has an important role for the synthesis of both gRNA and sgRNA.

A short 5′ UTR between 11 and 85 nt long precedes ORF1a. Also, an 80–85-nt long 3′ UTR is found between ORF2 and the poly(A) tail of HAstV genomes, but this sequence can be longer (130–305 nt) in avian astroviruses [68]. Of note, in addition to the 3′ UTR, the terminal 19 nt of ORF2 are also highly conserved among HAstV serotypes [48], and similarities have been observed in the sequence and folding of the 3′ UTR of several animal astroviruses, as well as in avian infectious bronchitis virus (a coronavirus), a dog norovirus, and the equine rhinovirus serotype 2 [36]. These conserved sequences are thought to be important for interaction with the viral RNA replicase and with cellular proteins that are likely needed for replication of the viral genome.

General Features of Nonstructural Proteins

The nonstructural proteins synthesized by positive-sense, single-stranded RNA viruses usually include a protease, an RNA-dependent RNA polymerase (RdRp), a helicase, and, in some cases, a viral genome-linked protein (VPg). The helicase and NTPase unwind the RNA before replication, while the VPg is covalently attached to the 5′-end of the RNA genome and is required for its efficient translation and/or replication. With the exception of the helicase, there is evidence for the presence of the other activities in the nonstructural proteins of astroviruses.

The primary translation products of the astrovirus genome are polyproteins nsp1a and nsp1ab (Fig. 2.3). Nsp1a is encoded in ORF1a, while nsp1ab is translated from ORF1a and ORF1b through a ribosomal frameshift mechanism. The length of nsp1a varies in mammalian viruses from 787 to 950 aa long, and in avian viruses it can be as long as 1,240 aa (Table 2.1). The most evident feature of nsp1a is the presence of a protease motif. The sequence of the predicted protease is well conserved between HAstV strains, but less conserved between mammalian and avian astroviruses, with only 37% amino acid identity between HAstV-1, turkey astrovirus, and avian necrosis virus [34]. The protein displays the typical properties of trypsin-like enzymes. The crystal structure (2.0 Å resolution) of the predicted protease from a HAstV strain revealed a similarity with viral serine proteases from Sesbania mosaic virus and various picornaviruses (human rhinovirus 2, hepatitis A

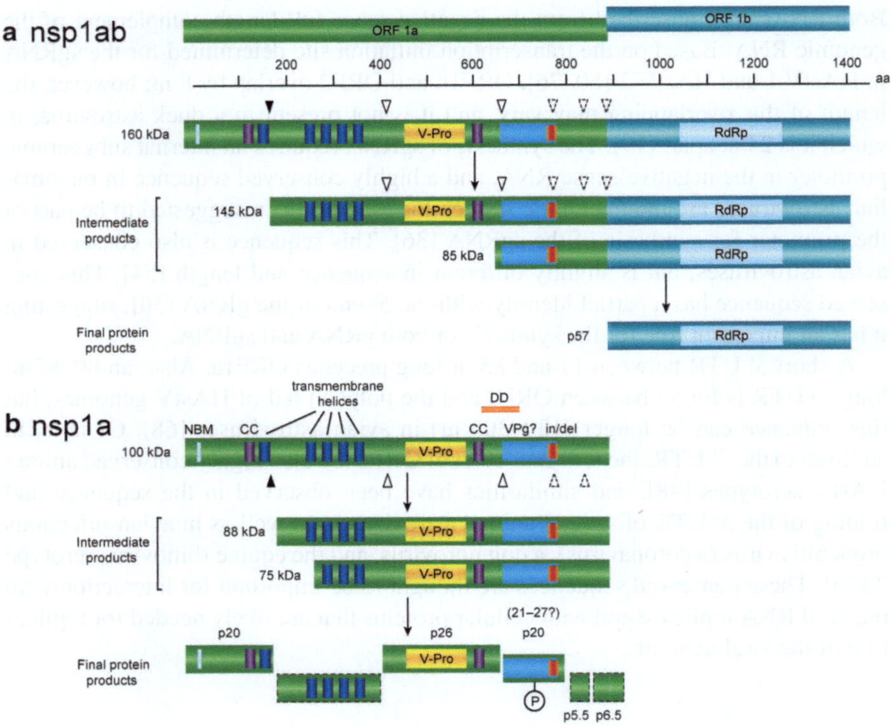

Fig. 2.3 Nonstructural proteins and protein structural motifs predicted from the nucleotide sequence of ORF1a and ORF1b. *NBM* nucleotide binding motif, *CC* coiled-coil, *TH* transmembrane helices, *v-Pro* viral protease, *VPg* viral protein attached to the genome, *in/del* insertion/deletion, *DD* putative death domain, *RdRp* RNA-dependent RNA polymerase. Nsp1ab (**a**) and nsp1a (**b**) are processed by cellular (*closed triangle*) and viral (*open triangles*) proteases. Although only the cleavages around amino acid residues 410 and 654 have been confirmed to be due to v-Pro, indirect evidences that would explain the protein processing products observed suggest that the downstream cleavages (*dashed triangles*) are also due to the viral protease. No products from the hydrophobic region (TH) have been identified, and the smaller products of 5.5 and 6.5 kDa have not been mapped into nsp1a (*dotted boxes*)

virus, and foot-and-mouth disease virus). The structural comparison confirmed that the catalytic triad is composed of His461, Asp489, and Ser551 [63], as was initially proposed using site-directed mutagenesis [37].

Sequence analysis of nsp1a showed the presence of five to six transmembrane helices upstream the protease motif, both in mammalian and avian viruses, although there is little conservation at the amino acid level between different astrovirus strains. Two regions with heptad repeat patterns were also identified in nsp1a as coiled-coil structures, one just upstream of the first helical motif and the second downstream of the protease motif [36]. The function of these structures or of the proteins that contain them remains unknown.

Table 2.3 Sequence diversity of protein nsp1b of astroviruses[a]

	Virus	H-1	H-3	H-4	H-5	H-6	H-8	H-MLB1	H-MLB2	H-SG	H-VA1	H-VA2	HMO-A	P-2	P-4	P-5	B	O-2	Rat	M	Csl-4	Csl-5	Csl-6	Csl-7	Csl-8	Csl-9	Csl-10	Csl-11	D	ANV-1	C	T	T-2
Human	HAstV-1	100																															
	HAstV-3	94	100																														
	HAstV-4	99	94	100																													
	HAstV-5	96	95	97	100																												
	HAstV-6	95	95	95	95	100																											
	HAstV-8	99	94	98	97	95	100																										
	HAstVMLB1	56	55	54	55	54	56	100																									
	HAstVMLB2	56	56	56	56	55	56	80	100																								
	HAstV-SG	52	53	52	53	53	52	48	50	100																							
	HAstV-VA1	52	53	52	52	52	51	48	49	98	100																						
	HAstV-VA2	51	51	51	51	50	51	48	49	74	75	100																					
	HMOAstV-A	52	52	52	52	51	52	48	48	74	75	97	100																				
Porcine	PAstV-2	57	58	58	56	58	57	53	54	52	54	51	52	100																			
	PAstV-4	57	58	58	56	58	57	53	54	51	54	51	52	60	100																		
	PAstV-5	49	49	49	49	48	49	47	47	47	47	44	45	49	49	100																	
Bovine	BAstV	61	61	61	61	61	62	54	56	51	52	50	51	83	61	51	100																
Ovine	OAstV-2	51	51	51	51	50	51	48	47	63	64	64	66	50	50	46	51	100															
Rat	RatAstV	58	57	57	56	57	57	57	56	49	49	49	50	58	58	47	59	48	100														
Mink	MAstV	51	51	51	51	51	51	47	48	64	64	65	65	53	52	46	53	64	51	100													
Sea lion	CslAstV-4	72	72	72	72	72	72	55	56	52	53	52	53	57	57	46	59	50	57	50	100												
	CslAstV-5	90	92	90	90	90	91	54	57	54	55	51	53	58	58	50	62	52	56	52	72	100											
	CslAstV-6	72	72	72	72	72	72	55	55	52	54	52	53	57	57	46	59	51	58	51	98	72	100										
	CslAstV-7	72	72	73	73	73	72	55	56	52	54	52	54	57	58	46	60	50	59	51	88	72	88	100									
	CslAstV-8	72	72	72	72	72	72	55	56	52	53	52	53	57	58	45	59	50	57	51	98	71	97	87	100								
	CslAstV-9	72	73	73	73	73	72	55	56	52	54	52	54	57	59	46	59	50	58	51	88	72	88	99	87	100							
	CslAstV-10	68	68	69	69	68	68	51	53	48	49	49	50	55	54	44	56	47	56	48	83	68	83	94	82	93	100						
	CslAstV-11	90	92	90	90	90	90	54	57	54	55	51	53	58	58	49	62	52	56	52	72	99	72	72	71	72	68	100					
Duck	DAstV	42	42	42	42	43	42	40	40	40	41	39	41	42	41	43	42	41	39	37	42	42	42	43	42	43	40	42	100				
	ANV-1	44	44	43	44	45	44	43	43	39	39	39	39	42	42	41	42	39	41	39	42	43	42	43	41	43	40	42	54	100			
	CAstV	41	42	41	41	43	41	40	42	41	42	42	44	40	41	41	42	40	40	41	42	42	43	44	42	44	40	42	70	55	100		
Turkey	TAstV	41	42	41	42	41	41	42	43	41	42	42	43	41	41	41	42	42	37	41	42	44	42	42	42	42	38	44	70	54	70	100	
	TAstV-2	42	41	41	42	41	42	40	42	41	43	41	43	41	40	40	42	42	38	41	42	43	42	41	41	41	38	43	70	54	70	95	100

[a]All available nsp1b sequences at the NCBI database were downloaded for each astrovirus strain. The sequences were pairwised compared, and their percent amino acid identities are shown in the figure.

Phylogenetic and comparative analysis of the genome of astrovirus with other plus-stranded RNA viruses identified a putative VPg coding region between amino acids 664 and 758 of nsp1a, with high sequence similarity to the VPg coding region of caliciviruses [2]. Its synthesis in astrovirus-infected cells has not been investigated; however, similar to calicivirus, treatment of the viral genomic RNA with proteases markedly reduced the yield of infectious virus when the astrovirus genome was transfected into permissive cells, suggesting that the ability of the astrovirus genome to initiate a replication cycle depends on the presence of a genome-linked protein [31, 70]. A hypervariable region with deletions and insertions downstream the VPg motif has been related to adaptation of the virus to culture cells [73] and to efficient synthesis of viral RNA during infection [30]. However, the form in which this region might modulate these events is unknown. The role in viral replication of the C-terminal region of nsp1a, comprising from aa residue 568 to the C terminus of the protein has been analyzed. This region, named nsp1a/4 [29], includes the putative VPg protein and the hypervariable region, and was found to be phosphorylated and to form oligomers. Nsp1a/4 interacts with the RNA polymerase when translated *in vitro*, and this interaction depends on the phosphorylation status of the protein. The fact that the proposed phosphorylation site is in the putative VPg region supports the idea that nsp1a/4 could be implicated in the replication of the astrovirus genome [20]. In addition, nsp1a/4 has a perinuclear distribution near the endoplasmic reticulum and colocalizes with viral RNA [29]. Whether the region of nsp1a/4 that interacts with the replicase is equivalent to the predicted VPg protein remains to be determined.

A bipartite nuclear localization signal (NLS) was described downstream of the second coiled-coil motif in HAstV and avian necrosis virus. When a recombinant protein spanning the NLS of HAstV-1 was expressed using baculovirus, the protein localized to the nucleus of insect cells [74]; however, the NLS antigen has not been observed in astrovirus-infected cells [24]. Moreover, neither turkey nor ovine astrovirus genome has a conserved NLS motif [36] and the region where the NLS motif is located has been proposed to encode the astrovirus VPg [2]. Thus, the role of the NLS remains unknown.

Motifs for an RNA-dependent RNA polymerase (RdRp) have been identified in the polypeptide encoded in ORF1b. This ORF encodes a protein of 515–539 amino acids that contains a region similar to the RdRp of plant bymoviruses and potyviruses [34, 43]. The eight conserved motifs in this region are typical of the RdRp of positive strand RNA viruses of the Koonin's supergroup I [39], which includes picornaviruses and caliciviruses (Table 2.3).

A region potentially encoding an RNA helicase has not been found in the astrovirus genome. This is unusual in single-stranded RNA virus with a genome length of more than 6,000 nt [34]. A conserved sequence in the N-terminal region of nsp1a of mammalian astrovirus has similarity with motifs I and II (NTP binding motif) of the seven conserved motifs of the RNA helicase of pestivirus, but lacks other motifs, such as the substrate binding and NTP hydrolysis domains present in these enzymes [67]. However, other similarities were not found, and this region is not conserved in avian astroviruses [2].

Table 2.4 Sequence diversity of the capsid polyprotein of astroviruses[a]

Host	Virus	H-1	H-2	H-3	H-4	H-5	H-6	H-7	H-8	H-MLB1	H-MLB2	H-SG	H-VA1	H-VA2	HMO-A	HMO-B	HMO-C	P	P-2	P-4	P-5	B	Cc-1	Cc2	O	O-2	Ca	SHO
Human	HAstV-1	100																										
	HAstV-2	66	100																									
	HAstV-3	82	71	100																								
	HAstV-4	63	65	65	100																							
	HAstV-5	70	66	72	64	100																						
	HAstV-6	70	67	72	65	75	100																					
	HAstV-7	70	67	82	61	71	69	100																				
	HAstV-8	71	70	73	71	74	72	71	100																			
	HAstVMLB1	40	38	38	32	38	29	32	30	100																		
	HAstVMLB2	41	30	31	29	29	39	31	31	74	100																	
	HAstV-SG	40	41	40	40	39	40	40	40	38	37	100																
	HAstV-VA1	37	41	41	41	40	40	41	41	37	36	95	100															
	HAstV-VA2	39	40	40	39	40	40	40	39	37	38	52	52	100														
	HMOAstV-A	40	41	40	40	40	40	39	39	37	38	49	50	99	100													
	HMOAstV-B	39	39	38	37	38	39	38	39	37	39	74	76	52	50	100												
	HMOAstV-C	37	42	41	41	40	40	41	42	37	36	95	99	52	50	76	100											
Porcine	PAstV	45	46	46	41	45	44	45	46	42	39	40	41	42	42	39	41	100										
	PAstV-2	39	39	40	39	40	40	40	40	36	38	36	36	37	37	37	36	43	100									
	PAstV-4	42	40	42	42	40	42	43	42	38	39	34	38	38	38	37	38	44	59	100								
	PAstV-5	41	42	41	40	40	42	39	40	33	34	36	37	36	36	36	37	37	38	37	100							
Bovine	BAstV	44	43	44	42	44	45	43	40	39	39	37	38	34	37	39	38	40	46	32	42	100						
Deer	CcAstV-1	44	31	44	43	44	42	43	30	38	40	38	38	36	36	36	38	41	34	32	37	69	100					
	CcAstV-2	44	43	44	43	43	42	44	43	40	40	35	35	37	37	34	35	41	46	32	38	69	68	100				
Ovine	OAstV	41	42	39	41	40	39	35	35	40	37	44	45	62	62	44	45	42	34	36	35	34	35	34	100			
	OAstV-2	41	43	42	40	40	43	39	41	35	35	38	36	36	36	34	36	37	33	36	77	35	35	37	37	100		
Canine	CaAstV	37	53	36	51	35	53	42	37	38	42	45	44	42	42	44	44	53	40	41	39	42	42	45	45	40	100	
	SHOAstV	37	50	41	49	37	37	36	39	36	35	42	43	42	42	42	43	53	42	38	38	40	41	24	41	40	76	100
Feline	FAstV	81	59	61	56	62	59	60	60	39	31	40	40	39	40	40	40	48	36	29	40	42	43	41	36	40	37	51
Bat	BatAstV-1	40	42	43	42	40	42	42	42	37	38	48	49	49	50	48	49	41	35	37	36	36	36	33	52	42	43	42
Mouse	MouseAstV	41	41	38	41	41	41	39	41	35	36	49	51	45	49	50	52	39	37	39	39	35	37	36	51	35	39	44
Rat	RatAstV	38	37	36	38	39	39	36	36	35	28	36	35	32	32	37	35	37	37	41	39	39	30	41	34	39	41	31
Mink	MAstV	35	36	35	36	39	36	35	36	35	36	42	41	46	42	41	41	39	36	36	35	30	33	31	53	35	37	37
Sea lion	CslAstV-1	36	38	39	39	39	38	36	34	37	36	45	44	47	43	43	44	40	37	33	35	33	35	34	44	35	40	37
	CslAstV-2	40	55	54	53	56	57	55	39	38	39	42	43	36	36	43	43	59	37	39	43	35	35	36	40	43	39	39
	CslAstV-4	38	54	38	55	39	40	53	41	39	42	44	44	42	42	44	44	41	36	43	40	35	38	37	41	40	39	39
	CslAstV-5	44	42	41	47	42	43	40	42	29	42	43	44	42	42	44	44	49	40	37	38	39	33	40	45	38	41	39
	CslAstV-6	58	60	58	59	59	58	59	42	39	43	44	42	41	41	40	42	40	37	41	39	37	40	39	41	40	38	43
	CslAstV-7	57	54	42	55	61	56	38	42	40	42	40	39	36	39	40	39	40	36	37	39	41	41	39	39	42	39	40
	CslAstV-8	57	56	57	42	58	57	57	56	39	41	41	40	41	41	40	40	54	38	41	40	36	39	37	38	39	40	41
	CslAstV-9	40	55	40	56	57	38	55	41	45	44	42	42	41	41	41	42	42	39	41	41	41	45	41	40	44	39	39
	CslAstV-10	40	56	56	55	40	40	55	57	38	41	41	40	40	41	40	40	55	37	39	39	35	38	37	40	38	53	39
	CslAstV-11	43	43	42	44	44	44	44	44	36	41	40	40	40	40	38	41	44	35	41	38	40	38	39	37	40	39	39
Dolphin	BdAstV-1	44	46	38	44	42	46	42	38	37	41	42	43	40	41	41	43	41	38	37	37	36	36	37	43	37	50	48
	BdAstV-3	40	38	40	40	40	39	42	40	35	38	38	40	37	36	38	40	38	36	34	39	33	37	38	36	39	39	40
Whale	MWAstV-2	42	42	40	42	42	42	40	41	41	42	40	41	41	41	40	41	31	40	38	40	32	36	36	41	41	43	42
Pigeon	FPAstV	31	31	31	30	31	31	31	31	27	26	30	30	31	30	31	30	29	26	28	29	26	26	25	29	30	31	31
	WPAstV	31	32	30	31	29	30	29	31	25	25	32	33	32	32	31	33	31	27	26	30	25	25	25	29	31	30	31
Duck	DAstV	25	25	26	27	27	26	26	27	28	28	31	30	30	30	32	30	25	26	23	29	24	23	27	32	28	30	30
Chicken	ANV-1	29	29	29	29	29	29	28	29	25	27	32	32	31	31	31	32	29	26	25	27	27	26	26	28	29	28	28
	ANV-2	31	33	31	33	30	32	32	33	27	26	29	30	27	28	32	30	30	24	24	29	24	27	26	31	29	28	28
	CAstV	30	30	34	31	32	30	30	31	28	28	30	31	32	31	31	31	30	25	29	29	31	28	30	33	29	31	29
Turkey	TAstV	26	26	25	26	26	25	26	27	30	30	31	32	31	33	30	32	27	25	25	28	24	26	26	34	27	28	27
	TAstV-2	25	25	26	26	27	25	26	26	32	33	33	32	30	30	31	32	27	25	25	28	24	26	26	34	27	28	27
	TAstV-3	25	25	25	27	26	26	26	27	30	32	29	32	29	29	30	32	26	24	23	30	24	25	26	31	28	29	28

Host	Virus	F	Bat-1	Mouse	Rat	M	Csl-1	Csl-2	Csl-4	Csl-5	Csl-6	Csl-7	Csl-8	Csl-9	Csl-10	Csl-11	Bd1	Bd-3	MW-2	FP	WP	D	ANV-1	ANV-2	C	T	T-2	T-3
Feline	FAstV	100																										
Bat	BatAstV-1	40	100																									
Mouse	MouseAstV	41	47	100																								
Rat	RatAstV	38	37	29	100																							
Mink	MAstV	35	49	44	29	100																						
Sea lion	CslAstV-1	39	47	45	33	63	100																					
	CslAstV-2	57	42	38	36	36	38	100																				
	CslAstV-4	54	42	39	37	40	39	57	100																			
	CslAstV-5	40	43	37	33	40	38	56	41	100																		
	CslAstV-6	60	43	42	37	38	37	55	56	42	100																	
	CslAstV-7	61	43	40	38	37	37	55	55	40	76	100																
	CslAstV-8	57	41	40	38	36	36	57	54	39	66	61	100															
	CslAstV-9	57	44	39	38	39	36	55	55	53	65	67	61	100														
	CslAstV-10	39	43	40	39	37	37	54	56	40	67	64	85	62	100													
	CslAstV-11	44	44	36	37	34	38	55	42	50	60	58	44	60	43	100												
Dolphin	BdAstV-1	44	41	37	35	40	40	47	46	44	49	47	48	43	48	45	100											
	BdAstV-3	40	41	39	36	37	30	41	41	41	41	39	40	42	41	40	40	100										
Whale	MWAstV-2	41	42	38	37	40	36	42	43	43	43	40	42	42	42	43	46	42	100									
Pigeon	FPAstV	31	27	29	26	28	26	30	31	30	30	29	29	29	29	29	31	32	34	100								
	WPAstV	31	28	27	28	30	31	28	31	32	30	31	29	30	28	29	28	33	31	55	100							
Duck	DAstV	25	28	29	27	28	26	26	26	27	27	27	29	27	29	27	28	28	27	37	36	100						
Chicken	ANV-1	28	25	28	28	28	28	29	27	31	27	29	26	30	25	28	27	32	28	63	49	35	100					
	ANV-2	31	27	29	27	27	27	30	30	30	27	32	31	33	31	30	28	32	31	63	63	31	60	100				
	CAstV	31	29	29	35	29	31	31	33	30	29	31	34	31	32	27	27	32	31	38	34	37	35	36	100			
Turkey	TAstV	25	28	30	28	28	32	27	28	25	27	27	26	27	27	26	29	28	31	35	37	70	36	33	43	100		
	TAstV-2	26	27	31	29	28	32	28	28	25	27	28	26	27	28	26	29	28	31	35	36	69	36	33	38	93	100	
	TAstV-3	27	27	30	28	26	30	27	25	25	27	26	27	29	26	26	30	28	30	39	36	72	35	33	40	84	83	100

[a]All available capsid amino acid sequences at the NCBI database were downloaded for each astrovirus strain. The sequences were pairwised compared, and their percent amino acid identities are shown in the figure.

General Features of Structural Proteins

The structural polyprotein shows the largest sequence variability among astrovirus proteins. It varies from 672 to 851 aa in length, and, in general, the ORF2 of avian viruses codes for shorter polyproteins (Table 2.4). The N-terminal half of the polyprotein is more conserved than the C-terminal domain, with the N-terminal region encompassing a region of basic amino acids that is conserved among astrovirus [35, 72] and is thought to interact with the genomic RNA in the virion [23]. The C-terminal half of the protein shows considerable sequence variability among astroviruses isolated from different animal species, or even among different isolates from the same species (Fig. 2.4) as it contains a delimited region with abundant

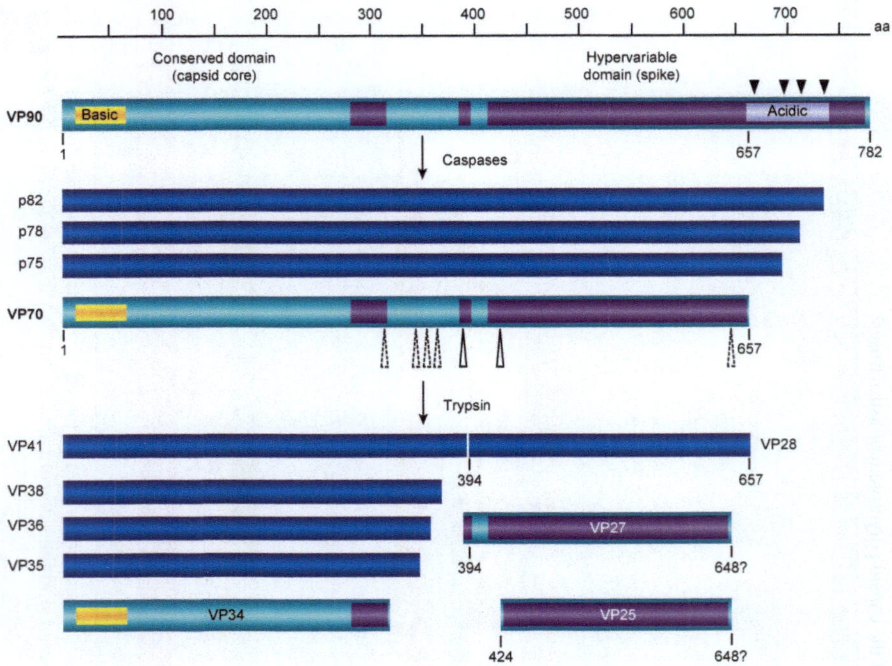

Fig. 2.4 Features of structural proteins. The primary ORF2 product (named VP90 in HAstV-8) contains two domains that can be distinguished by their degree of conservation: the N-terminal domain is highly conserved and forms the core of the capsid, while the hypervariable C-terminal domain forms the spikes of the virus particle. VP90 contains basic and acidic regions that are highly conserved among all astroviruses. VP90, the primary product of ORF2 is sequentially processed at its carboxy terminus by caspases to generate VP70 (intermediate cleavage products are shown in *gray*), the protein present in the extracellular virions. These particles are processed by trypsin to generate protein intermediates of variable size (in *gray*), and the final products VP34, VP27, and VP25. *Closed arrow heads* represent cleavages carried out by caspases, important for virus maturation and cell egress; *open long arrow heads*, cleavages by trypsin. The cleavage sites indicated by *dashed arrow heads* have not been precisely determined; therefore, the carboxy termini of the intermediate and final products are unknown

insertion or deletions [72]. The carboxy terminus of the polyprotein has an acidic character (residues 649–702 for HAstV-8) that is proposed to be highly conserved among all members of the *Astroviridae* family (Fig. 2.4) [54] and is the substrate of cellular caspases. The last five amino acid residues at the C-terminal end are also conserved among human and other mammalian virus strains. The conserved N-terminal domain of the structural protein forms the capsid core of the particle, whereas the hypervariable C-terminal domain forms the spikes of the virion. Thus, this last domain is predicted to participate in the early interactions of the virus with the host cell [16, 40].

Early Steps of Infection

The early interactions of astroviruses with their host cell have been poorly characterized. The cellular receptor is unknown; however, the observation that different HAstV serotypes show different tropism in cultured cells [7] suggests that more than one cellular attachment or receptor molecule may be used by these viruses. Caco-2 cells are susceptible to infection by all HAstV serotypes, but HTC-15, another colon carcinoma cell line, only weakly supports infection by HAstV-1 [7]. Likewise, baby hamster kidney (BHK-21) cells only permit the sustained passage of HAstV-2, but they are permissive for replication of other HAstV serotypes after transfection of *in vitro* transcribed genomic RNA [21], suggesting that the blockage is at the virus entry level. It has been found that HAstV-8 binds to the surface of Caco-2 cells with a half-time of about 9 min (Méndez et al., unpublished), and it is known that the infectivity of HAstV is greatly enhanced (3–5 logs) by, and probably is dependent on, trypsin cleavage of the capsid protein [4, 55, 61]. Although the trypsin cleavage products have been characterized (see below), the mechanism by which this treatment enhances virus infectivity is unknown. Trypsin cleavage of the precursor polyprotein produces in HAstV-8 three final products, VP34, VP25, and VP27 [55], with the concomitant induction of drastic structural changes in the particles [4]. VP34 represents the conserved domain and forms the capsid core, while both VP25 and VP27 form the spikes on the virion surface. The crystal structure of the HAstV-8 spike domain was recently solved at a 1.8 Å resolution [16], mapping conserved residues onto the astrovirus projection domain revealed a putative receptor binding site with amino acid compositions characteristic for polysaccharide recognition [16]. Antibodies that recognize the spike proteins VP25 and VP27 are serotype specific and neutralize virus infectivity [5, 61], probably by blocking virus binding.

Early studies with HAstV-1 in HEK293 Graham cells demonstrated the presence of virions in coated pits and coated vesicles, implicating clathrin-mediated endocytosis as the viral entry mechanism. This was supported by inhibition of viral infection by endocytosis-blocking drugs [15]. An acidic endosomal pH was also found to be required for HAstV-1 infection [15]. Recently, a clathrin-dependent endocytosis was confirmed to be a functional pathway for the entry of HAstV-8 into Caco-2 cells. Drugs that affect clathrin assembly, endosome acidification, and actin filaments, as well as those that decrease cellular cholesterol and interfere with vesicular transport, reduced the infectivity of HAstV-8. The infection of this virus was also reduced by knocking down the expression of clathrin heavy chain by RNA interference and by expression of dominant-negative mutants of dynamin (Méndez et al., unpublished data).

The mechanism through which the astrovirus genome is released from the infecting virus particle into the cytoplasm for translation, the cell site where it occurs, and the cellular and viral factors involved in this event are unknown. However, it has been determined that in Caco-2 cells, HAstV-8 uncoats about 130 min after cell entry, and the infectious virus has to reach late endosomes for a productive infection, since virus infectivity is reduced by knocking down the expression of Rab7 (Méndez et al., unpublished data).

The interaction of HAstV with the host cell provokes activation of the extracellular signal-regulated kinase (ERK1/2) pathway within the first 15 min after virus attachment. Although the mechanism for this activation is unknown, it was independent of virus replication, suggesting that this pathway is triggered during virus binding or entry into the cell [51]. ERK1/2 seems to be required at early times to establish a productive infection, since inhibitors of this kinase blocked synthesis of viral proteins and RNA, and, consequently, reduced the virus yield. It has also been shown that astrovirus increases barrier permeability in polarized monolayers of Caco-2 cells following apical infection [52]. Permeability was increased when monolayers were treated not only with infectious virus, but also with inactivated virus or purified virus-like particles (VLPs), indicating that astrovirus-induced permeability occurs independently of viral replication and is modulated by the capsid protein. It has been suggested that HAstV could trigger tight junction instability to reach putative receptors present in the basolateral membrane; however, the fact that the increase in permeability is observed at least 16 h postinfection suggests that more than a requirement for virus entry, it is a consequence of the cellular signal transduction pathways induced by the virus.

Translation

After uncoating, the virus RNA genome is translated into nonstructural proteins. It has been suggested that the astrovirus genome contains a VPg 5′ moiety in place of a traditional 7-methyl cap, and, as observed for caliciviruses, it could recruit the cap binding protein eIF4E to initiate translation [2, 10, 70]. The VPg protein from calicivirus has been shown to directly interact with eIF4E and eIF3, a multi-subunit scaffolding initiation factor, to promote assembly of the preinitiation complex [13, 25]. This VPg-eIF4E interaction may be a preferential method of viral over host translation [60].

Nonstructural Polyprotein Synthesis and Processing

The nonstructural astrovirus proteins are translated as two polyproteins, nsp1a (~101 kDa) and nsp1ab (~160 kDa). The translation of ORF1a to produce nsp1a stops at an amber codon, while a translational fusion of ORF1a with ORF1b that results from a − 1 frameshift at the junction of these two ORFs and leads to the synthesis of nsp1ab. Nsp1ab is translated between 7% (*in vitro*) and 25% (*in vivo*) of the time [34, 44] and requires a highly conserved heptameric "slippery" sequence (AAAAAAC) and a downstream hairpin structure for induction of the frameshift (Fig. 2.2) [43, 75]. Of note, mechanisms for attenuated translation of the RdRp are common in positive-sense RNA viruses [46].

The nsp1a and nsp1ab polyproteins are cleaved by the viral serine protease to produce the individual nonstructural proteins, although a cellular protease seems to be involved in at least one of the cleavage events [23, 24, 37, 56]. The number and size of the final nonstructural proteins is not fully elucidated, and studies using different cell lines, human virus serotypes, and expression systems have described different intermediate and final processing products [54]. However, conclusions drawn from these reports suggest that at least five nonstructural proteins exist (Fig. 2.3) (1) an N-terminal 20-kDa protein, likely cleaved co-translationally by a cellular signalase that contains one predicted transmembrane domain and a coiled-coil domain [56]; (2) a protein predicted to contain four transmembrane domains, although a protein from this hydrophobic region of nsp1a has not been identified; (3) a 27-kDa serine protease [22, 37]; (4) a phosphoprotein of 21–27 kDa depending on its level of phosphorylation or differential processing that contains the hypervariable nsp1a region and possibly is the VPg protein [2, 20, 29]; and (5) a 57-kDa RNA-dependent RNA polymerase [22, 43]. Finally, two reported small products of 5.5 and 6.5 kDa have not been mapped in nsp1a [74].

Structural Polyprotein Synthesis and Processing

The primary translation product of astrovirus ORF2 is an 87–90 kDa polyprotein synthesized from the viral sgRNA. The mechanism of translation of the sgRNA remains to be determined; however, as proposed for the astrovirus gRNA, it could be translated through a mechanism involving the presence of a 5′-end VPg, as it has been shown for the gRNA and sgRNA of caliciviruses [31, 70].

In HAstV-8-infected Caco-2 cells, a 90-kDa protein (VP90) is synthesized from ORF2. This protein assembles into viral particles and is then cleaved at the carboxy terminus by cellular caspases to produce a 70-kDa protein (VP70) through several intermediate cleavage events (Fig. 2.4) [55]. The VP90 to VP70 cleavage is associated with the exit of the virus from the cell. A 90 kDa protein in HAstV-1 infected cells has also been detected [49, 70]. The intracellular cleavage of the primary translation product of ORF2 still requires further investigation; however, it seems clear that extracellular virions are only constituted by VP70 [55]. These virions require trypsin treatment to fully activate their infectivity. The proteolytic treatment cleaves VP70 into three smaller products and enhances virus infectivity up to 105-fold [4, 42]. The trypsin cleavage pathway of VP70 has been best characterized for HAstV-8 (Fig. 2.4) and yields cleavage products of 25, 27, and 34 kDa. The cleavage products are similar in other astrovirus strains studied, although their sizes range from 24 to 26, 27 to 29, and 32 to 34 kDa [4, 6, 12, 23, 55]. Of interest, if intracellular viral particles formed by VP90 are isolated and treated with trypsin *in vitro* they also become infectious [53].

Transcription/Replication

Genome replication of positive-sense, single-stranded RNA viruses belonging to several different families has been well documented. Few details are known about astrovirus, but the available information is consistent with previous studies in other positive-sense RNA viruses, making it reasonable to assume that replication of the astrovirus genome occurs in a similar manner. Based on this premise, it is thought that after translation of the gRNA and proteolytic processing of the nonstructural proteins a viral replicase complex is formed, which uses the gRNA to produce a full-length negative-sense RNA. This negative-sense RNA is detected 6–12 h postinfection, and it accumulates to 0.7–4% of gRNA [33]. The replicase complex of positive-sense RNA viruses usually contains the viral RdRp and other nonstructural proteins, as well as host proteins necessary for replication of the viral genome. In the case of astrovirus, the nonstructural protein nsp1a/4 protein (corresponding to the C-terminal end of nsp1a) has been shown to be required for virus replication in addition to the viral polymerase [20, 30]. Also, a predicted VPg has been suggested as a possible product of nsp1a [2], opening the possibility that astrovirus RNA synthesis could be primer dependent. However, further studies are required to confirm this hypothesis. Whether the nonstructural proteins other than nsp1a/4 are required for astrovirus genome replication is unknown. Transcription of gRNA and sgRNA from an internal promoter of the full-length negative-sense RNA is detected starting at 8 h postinfection [54]. Requirements for the replicase complex to switch from minus-strand synthesis to positive-sense and sgRNA synthesis are unknown, but could pattern alphaviruses, which use nonstructural polyprotein processing to direct template activities [65].

The sgRNA is transcribed at an estimated tenfold greater amount over viral genome at 12 h postinfection [49]. Delayed structural protein synthesis and transcription of large amounts of sgRNA is a common theme in positive-sense RNA viruses and has been suggested to be a mechanism to overcome host translational repression, and possibly to promote packaging into virions of the newly synthesized viral genomes through inhibiting their translation [66]. The genome of HAstV contains highly conserved 5′ and 3′ UTRs [48]. UTRs from other viruses have been shown to facilitate 5′ and 3′ interactions during gRNA replication, recruit host factors for replication, are sites of recognition and initiation by the viral polymerase, and enhance the stability of the RNA genome. It is likely that astrovirus UTRs play a similar role in the virus gRNA replication, although this remains to be determined [45, 59, 64, 69, 71]. Very little information is available regarding specific host proteins involved in replication of the astrovirus genome, although this process could require several RNA binding proteins, as reported for other plus-strand RNA viruses [1, 8, 11].

Infectious cDNA Clone

As a positive-sense RNA virus, the astrovirus genome is able to initiate a productive infection [21]. *In vitro* transcribed RNA from a full-length cDNA clone of the HAstV-1 genome has also proved to be infectious [21]. Using this clone, a reverse genetics system for HAstV-1 was established and used to analyze specific mutations in the structural [23] and nonstructural proteins [30]. This system consists of the *in vitro* transcribed RNA, which is transfected into the highly transfectable, but nonsusceptible and poorly permissive BHK-21 cells, harvesting the virus produced, and propagating this virus in the astrovirus-susceptible Caco-2 cell line. This protocol, although efficient in terms of virus yield after amplification in Caco-2 cells, has the drawback that the mutant virus is finally obtained after several rounds of replication in the latter cells, which might induce accumulation of additional mutations in the viral progeny. Thus, analysis of the phenotype associated with specific mutations introduced in the cDNA clone may be disguised. In fact, viruses with changes not related to the mutations originally introduced in the nonstructural region of HAstV-1 were found after virus recovery by this method [30]. More recently, a protocol that improves the reverse genetics system of astrovirus has been described [70]. It uses the highly transfectable and HAstV-susceptible and highly permissive hepatocyte cell line Huh7.5.1 making it possible to obtain higher titer virus yield in a single replication cycle, decreasing the frequency of potentially undesired mutations [70].

Cellular Sites of Viral Genome Replication

The replication of all positive-sense RNA viruses characterized takes place in association with membranes [14, 47, 58]. Infection with these viruses promotes intracellular membrane rearrangement yielding networked structures of vesicles, single- and double-membrane vesicles, and often invaginations or spherules that are associated with the synthesis of viral RNA and proteins, and with the assembly of virus particles [14]. The induced organelle-like structures can be derived from the ER, Golgi apparatus, and lysosomal or endosomal membranes, and the viral RNA and proteins, as well as the cellular proteins involved in these processes, have been found to form replication complex at these sites [14].

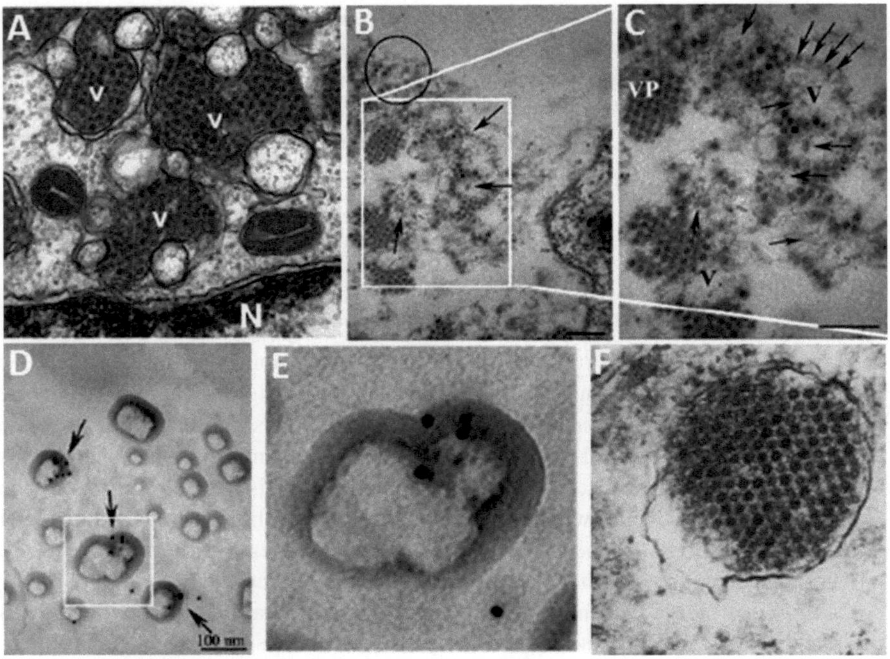

Fig. 2.5 Astrovirus proteins and particles are associated with membranes in infected cells. Ultrastructural analysis of Caco-2 cells infected with (**a**) HAstV-4 (48 hpi; *N* nucleus, *V* viral aggregates); (**b** and **c**) HAstV-8 (24 hpi; *V* vesicles, *VP* viral particles); (**d**) immunoelectron microscopy of membrane-enriched density gradient fractions of Caco-2 cells infected with HAstV-8. *Arrows* indicate positive immunogold signals; (**e**) enlarged image of (**d**); (**f**) electron microscopy of lamb mid-gut cells at 23 h postinfection with lamb astrovirus. Crystalline array of hollow-cored particles enclosed by a membrane. Photographs were reproduced from Guix et al. [29] (**a**), Mendez et al. [53] (**b–d**, **e**), and Gray et al. [27] (**f**), with permission

The membrane rearrangements induced by astrovirus infection have been poorly characterized. Viral aggregates in HAstV-4-infected cells were observed at 48 h postinfection near the periphery of the nucleus, surrounded by a large number of double-membrane vacuoles that were not observed in noninfected cells (Fig. 2.5a) [29]. Similar structures were found in HAstV-8-infected Caco-2 cells, where large groups of viral particles surrounding "O-ring" structures that probably correspond to double-membrane vesicles were described (Fig. 2.5b, c) [53]. Virus-like particles were observed inside and at the borders of the "O-ring" and could represent subviral particles at different assembling stages [53]. Moreover, the VP90 structural precursor protein was detected in membrane fractions resolved by density gradient centrifugation and analyzed by immunoelectron microscopy (Fig. 2.5d, e) [53]. Of interest, ultrastructural analysis of the jejunum and ileum of astrovirus-infected lambs showed viral particles aggregates within lysosomes and autophagic vacuoles of epithelial cells, as well as along microvilli (Fig. 2.5f) [27], suggesting that astrovirus replication associated with membranes also occurs in naturally targeted intestinal epithelial cells.

Structural and nonstructural astrovirus proteins associate with cellular membranes. The structural polyprotein of HAstV-8 interacts with membranous structures through its C-terminal region, since this region is protected from trypsin cleavage when VP90 is associated to membranes, but not when the protein is membrane free [53] and this interaction can occur in the absence of other viral proteins, as judged from *in vitro* translation assays [53].

The nsp1a/4 nonstructural protein of HAstV-4, which interacts with the RdRp and is involved in viral gRNA replication, has been reported to associate to the ER membrane in close vicinity to viral RNA and structural proteins [20, 29]. The protease and RdRp, as well as genomic and antigenomic (negative-sense) RNAs, have also been reported to associate with membranes in HAstV-8-infected cells [53]. The presence of structural (VP90) and nonstructural (nsp1a/4, protease, RdRp) viral proteins, as well as viral RNAs, in membranous fractions suggests that RNA replication and the first steps of virus morphogenesis are carried out in association with the observed membranous structures. It is still unknown, however, if RNA replication and virus assembly occur at the same subcellular membranous structure. The putative replicase complex of astrovirus seems to associate with membranes that probably derive from the ER, since viral phosphoproteins of 21–27 kDa, which interact with the RdRp [20], and are likely involved in RNA replication, localize to this organelle [29].

Virus Morphogenesis and Release

The assembly of viral particles is a highly complex process. This process requires the coordinated expression of genes encoding the structural proteins, the proper interaction of the proteins among themselves and with the virus genetic material, and in many instances coupling of the capsid assembly with the replication of the viral genome. In addition, the assembly process frequently occurs in association with cell membranes and with the help of cellular proteins. The knowledge on the assembly process of infectious astrovirus particles is limited, but it is known that the capsid assembles in the absence of viral RNA, carried out in membranous structures, and requires the activity of cellular caspases during the final assembly step.

Assembly of Viral Particles

Virus-like particles (VLPs) spontaneously assemble both in mammalian and insect cells when astrovirus ORF2 is expressed using heterologous systems. The expression of HAstV-2 ORF2 in LLCMK2 cells using a recombinant vaccinia virus led to the assembly of VLPs similar in size and morphology to native

HAstV-2 virus particles when analyzed by electron microscopy [12]. Moreover, when these VLPs were purified and treated with trypsin, proteins VP32, VP29, and VP26 were generated, in coincidence with the structural proteins found in native viral particles [12], indicating that the HAstV-2 capsids can assemble correctly in the absence of viral RNA.

VLPs were also obtained when the complete ORF2 of HAstV-1 was expressed using the baculovirus–insect cells system [9], although the particles presented an atypical morphology when purified [54]. VLPs were assembled in this system when a truncated form of the structural polyprotein lacking the N-terminal 70 amino acids was expressed, or when a protein whose 70 N-terminal amino acids had been replaced by the green fluorescent protein was synthesized [9], indicating that the 70 N-terminal amino acids of the structural polyprotein are not essential for virus particle assembly; although it is believed that this region, of basic nature, is important for interaction with the viral genome during morphogenesis of the viral particle [54]. When the VLPs assembled in insect cells were further characterized by ultracentrifugation in sucrose gradients, the particles banded at a density of 1.08 g/ml, and electron microscopy analysis showed the presence in this fraction of VLPs around 38 nm in diameter, similar to HAstV-1 purified particles, as well as ring-like structures of 16 nm in diameter [9]. Of interest, these ring-like structures could be converted into VLPs in the presence of Mg^{2+}, while sequestration of divalent cations promoted the disassembly of VLPs into ring-like structures [9]. These observations showed the importance of divalent ions to maintain the capsid structure and suggest that the ring-like structures could be directly involved in VLP morphogenesis. The ring-like structures were also observed when native HAstV-1 viruses were purified by CsCl centrifugation in the presence of EDTA.

The assembly of gRNA into viral particles requires the specific interaction of structural protein(s) with encapsidation signals on the viral RNA and, in the case of positive-sense RNA viruses, the process of particle assembly is thought to occur through condensation of the capsid protein(s) around the nucleic acid. In this context, it is interesting that VP90 has been observed to form oligomers (dimers and trimers) in virus-infected Caco-2 cells, suggesting that this protein can assemble into intermediary structures during virion assembly [53]. Moreover, the N-terminal region of VP90 has a basic character that is conserved among astrovirus and is believed to interact with gRNA during packaging and assembly [54]; however, there is no information yet about encapsidation signals in the astrovirus gRNA.

Role of Caspases

Although there are no reports exploring the role in virus assembly of the C-terminal region of the structural polyprotein, it is known that it has to be cleaved by caspases to yield the mature virion and to promote egress of the virus. Analysis of HAstV-8-infected cells by density ultracentrifugation showed that intracellular viral particles

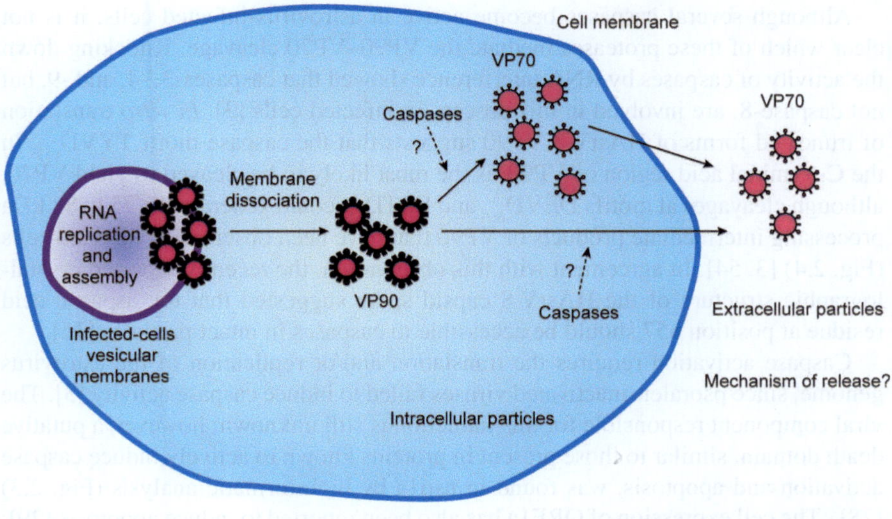

Fig. 2.6 Model of astrovirus particle maturation and cell egress. See details in the text. Modified from Méndez and Arias [53]

consisting of VP90 were present in association with membranes as well as in membrane-free fractions [53]. Based on these and related findings [3], a model was proposed in which the viral progeny assemble in membranous structures, and then dissociate from the membranes by an unknown mechanism; once in a membrane-free environment the assembled VP90 is then proteolytically cleaved at the C-terminal end to yield viral particles containing VP70 (Fig. 2.6). This cleavage is blocked by caspase inhibitors and induced by TRAIL (TNF-related apoptosis ligand), suggesting an active role of caspases in the process [57], although it is not clear if VP90 processing occurs intracellularly or concomitantly with cell egress of the virus. The involvement of caspases in this process is supported by the presence of an acidic region in the C-terminal region of VP90, conserved in astroviruses, that contains several putative caspases recognition motifs [57].

It has been reported that caspases-3 and -6 were activated during HAstV-8 infection of Caco-2 cells, as judged by cleavage of their respective substrates lamin A and PARP (poly ADP-ribose polymerase). The cleavage of PARP occurred at 12 h postinfection, in coincidence with the cleavage of VP90 to VP70 [3, 57]. Furthermore, executioner caspases-3/7, as well as initiator caspases-4, -8, and -9, were all activated at 12 h postinfection [3] as determined by the cleavage of preferential substrates of these caspases [3]. In addition, the precursor forms of all evaluated caspases were processed in these cells, confirming their active status, and suggesting that more than one caspase is active during astrovirus infection [3]. In contrast with these results, neither the apoptosis mediated by the mitochondrial pathway nor caspase-9 was activated in HAstV-4-infected Caco-2 cells, although the processed product of pro-caspase-8 was detected, suggesting its activation [28].

Although several caspases become active in astrovirus-infected cells, it is not clear which of these proteases mediate the VP90–VP70 cleavage. Knocking down the activity of caspases by RNA interference showed that caspases-3, -4, and -9, but not caspase-8, are involved in this process in infected cells [3]. *In vitro* translation of truncated forms of HAstV-8 VP90 suggests that the caspase motif $TYVD_{657}$, in the C-terminal acid region of VP90, is the most likely to be cleaved to yield VP70, although cleavages at motifs $DEVD_{666}$ and $EETD_{672}$ could generate the 75 to 82 kDa processing intermediate products of VP90 that have been observed in infected cells (Fig. 2.4) [3, 54]. In agreement with this observation, the recently reported crystallographic structure of the HAstV-8 capsid spike suggested that the aspartic acid residue at position 657 should be accessible to caspases in intact particles [16].

Caspase activation requires the translation and/or replication of the astrovirus genome, since psoralen-inactivated viruses failed to induce caspase activity [3]. The viral component responsible for this induction is still unknown; however, a putative death domain, similar to those present in proteins known to actively induce caspase activation and apoptosis, was found in nsp1a by bioinformatic analysis (Fig. 2.3) [28] . The cell expression of ORF1a has also been reported to induce apoptosis [29]; whether this effect is connected with the presence of the proposed death domain needs further investigation. Processing of viral proteins by caspases that result in the pro- or antiapoptotic activities of the processed viral polypeptides has been reported for other viruses such as influenza A and B [78], hepatitis C [62], and coronavirus [17]. However, astrovirus is one of the few viruses described in which caspase cleavage of viral proteins is required for the life cycle of the virus.

Virus Egress

The assembly of viral particles whose infectivity can be activated by *in vitro* treatment with trypsin starts after 8 h postinfection, reaching the highest level at 12 h postinfection (Fig. 2.7). The number of viral particles increased both intracellularly and in the cell medium. At 12 h postinfection only about 10% of the total particles were found extracellularly [3], and at 24 h postinfection the released virus accounted for about 30–50% of total viruses. The VP90–VP70 processing is increased by treatment of cells with the apoptosis inducer TRAIL, and concomitantly with this enhanced cleavage, a four to fivefold increase in the virus released to the cell medium was observed [57].

Although the cellular process triggered by caspase cleavage of VP90 to allow virus exit is unknown, it seems that cell death is neither induced by, nor required for this process, and rather a nonlytic mechanism seems to be involved [3]. Caspase activation does not always result in cell death or secondary necrosis and the activation level of these proteases and their interaction with cellular proteins modulate cell death progression [41]. The requirement of VP90 processing by caspases for virus exit has been confirmed through the use of caspase inhibitors. A potent inhibitor for executioner caspases-3/7, Ac-DEVD-CHO, blocked virus release at levels similar to

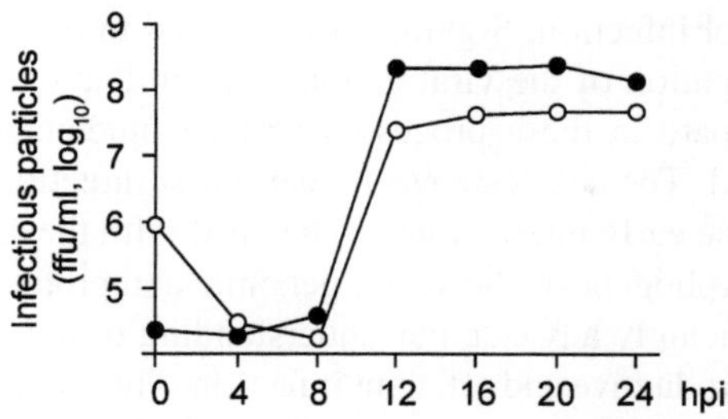

Fig. 2.7 Kinetics of infectious virus production. Caco-2 cells were infected with HAstV-8, and at different time points (from 0 to 24 hpi) the extracellular and cell-associated fractions were analyzed for infectious viral particles. Virus in the extracellular (*empty circles*) and cell-associated (*filled circles*) fractions was determined after trypsin treatment [3]

those induced by the pan-caspase inhibitor Z-VAD-fmk. On the other hand, inhibitors of initiator caspases-8 and -9 had no effect on virus egress [3]. Thus, although VP90 can be cleaved in astrovirus-infected cells by caspases-3 and -9, as described earlier, virus exit seems to depend on the activity of caspases 3/7 but not of caspase-9, and most probably on the activity of caspase-3, since RNA interference of the expression of this caspase has a strong negative effect on virus release [3].

Previous reports have associated the activity of executioner caspases with the intracellular transport of viral proteins, as in the case of influenza virus, where the activity of caspase-3 was required for the intracellular transport of NP protein [77]. The release of nonenveloped viruses, such as astrovirus, is generally believed to occur by cell membrane destabilization. However, rotavirus and poliovirus exit the cells through a nonlytic mechanism, and in the later case, probably through autophagic vesicles [32]. Thus, intracellular transport of astrovirus particles could be important for cell exit. It has not been determined whether autophagy is induced during astrovirus infection, although double membrane vesicles have been observed in astrovirus-infected cells [26, 29]. Some astrovirus strains have been shown to cause cytopathic effect on cultured cells and necrosis of hepatic cells in natural infections [26]. It is unclear which astrovirus-cell interaction leads to virus egress via a lytic or nonlytic process.

Perspectives

The perception of the relevance of astroviruses in public health is changing with the recent discovery of novel viruses isolated from humans that show genetic similarities with animal strains, some of which have been associated with diseases other than gastroenteritis. The continued application of high-resolution structural cryo-electron microscopy and X-ray crystallography of the virus and viral proteins will allow us to gain insight into the different stages of the life cycle of these viruses. Identification of the cell receptors and the entry pathway used by astroviruses to infect susceptible cells in culture, but most importantly to infect the target cells in a natural infection, will be of utmost importance to understand the biology of these viruses and to control infection. Significant gaps exist in our understanding of the translation and replication of the viral genome; identification of viral and cellular proteins that participate in these processes will be fundamental to elucidate the mechanisms involved. The cell response to astrovirus infection has proved already to be beneficial for the early interactions of the virus with the cell, as well as for the last step of virus morphogenesis; however, genomic and proteomic approaches will be required to significantly advance our understanding of the virus–host cell interactions that lead to productive and efficient infection. The recently improved reverse genetics system developed for astroviruses will also be key to identify the molecular determinants involved in the virus replication cycle.

References

1. Ahlquist P, Noueiry AO, Lee WM, Kushner DB, Dye BT. Host factors in positive-strand RNA virus genome replication. J Virol. 2003;77(15):8181–6.
2. Al-Mutairy B, Walter JE, Pothen A, Mitchell DK. Genome prediction of putative genome-linked viral protein (VPg) of astroviruses. Virus Genes. 2005;31(1):21–30.
3. Banos-Lara MDR, Méndez E. Role of individual caspases induced by astrovirus on the processing of its structural protein and its release from the cell through a non-lytic mechanism. Virology. 2010;401(2):322–32.
4. Bass DM, Qiu S. Proteolytic processing of the astrovirus capsid. J Virol. 2000;74(4):1810–4.
5. Bass DM, Upadhyayula U. Characterization of human serotype 1 astrovirus-neutralizing epitopes. J Virol. 1997;71(11):8666–71.
6. Belliot G, Laveran H, Monroe SS. Capsid protein composition of reference strains and wild isolates of human astroviruses. Virus Res. 1997;49(1):49–57.
7. Brinker JP, Blacklow NR, Herrmann JE. Human astrovirus isolation and propagation in multiple cell lines. Arch Virol. 2000;145(9):1847–56.
8. Burnham AJ, Gong L, Hardy RW. Heterogeneous nuclear ribonuclear protein K interacts with Sindbis virus nonstructural proteins and viral subgenomic mRNA. Virology. 2007;367(1):212–21.
9. Caballero S, Guix S, Ribes E, Bosch A, Pinto RM. Structural requirements of astrovirus virus-like particles assembled in insect cells. J Virol. 2004;78(23):13285–92.
10. Carter MJ, Willcocks MM. The molecular biology of astroviruses. Arch Virol Suppl. 1996; 12:277–85.
11. Cristea IM, Carroll JW, Rout MP, Rice CM, Chait BT, MacDonald MR. Tracking and elucidating alphavirus-host protein interactions. J Biol Chem. 2006;281(40):30269–78.
12. Dalton RM, Pastrana EP, Sánchez-Fauquier A. Vaccinia virus recombinant expressing an 87-kilodalton polyprotein that is sufficient to form astrovirus-like particles. J Virol. 2003; 77(16):9094–8.
13. Daughenbaugh KF, Fraser CS, Hershey JWB, Hardy ME. The genome-linked protein VPg of the Norwalk virus binds eIF3, suggesting its role in translation initiation complex recruitment. EMBO J. 2003;22(11):2852–9.
14. Denison MR. Seeking membranes: positive-strand RNA virus replication complexes. PLoS Biol. 2008;6(10):e270.
15. Donelli G, Superti F, Tinari A, Marziano ML. Mechanism of astrovirus entry into Graham 293 cells. J Med Virol. 1992;38(4):271–7.
16. Dong J, Dong L, Méndez E, Tao Y. Crystal structure of the human astrovirus capsid spike. Proc Natl Acad Sci USA. 2011;108(31):12681–6.
17. Eléouët JF, Slee E, Saurini F, Castagné N, Poncet D, Garrido C, et al. The viral nucleocapsid protein of transmissible gastroenteritis coronavirus (TGEV) is cleaved by caspase-6 and -7 during TGEV-induced apoptosis. J Virol. 2000;74(9):3975–83.

18. Firth AE, Atkins JF. Candidates in astroviruses, seadornaviruses, cytorhabdoviruses and coronaviruses for +1 frame overlapping genes accessed by leaky scanning. Virology. 2010; 7:17.
19. Fu Y, Pan M, Wang X, Xu Y, Xie X, Knowles NJ, et al. Complete sequence of a duck astrovirus associated with fatal hepatitis in ducklings. J Gen Virol. 2009;90(Pt 5):1104–8.
20. Fuentes C, Guix S, Bosch A, Pintó RM. The C-terminal nsP1a protein of human astrovirus is a phosphoprotein that interacts with the viral polymerase. J Virol. 2011;85(9):4470–9.
21. Geigenmüller U, Ginzton NH, Matsui SM. Construction of a genome-length cDNA clone for human astrovirus serotype 1 and synthesis of infectious RNA transcripts. J Virol. 1997;71(2): 1713–7.
22. Geigenmüller U, Chew T, Ginzton N, Matsui SM. Processing of nonstructural protein 1a of human astrovirus. J Virol. 2002;76(4):2003–8.
23. Geigenmüller U, Ginzton NH, Matsui SM. Studies on intracellular processing of the capsid protein of human astrovirus serotype 1 in infected cells. J Gen Virol. 2002;83(Pt 7):1691–5.
24. Gibson CA, Chen J, Monroe SA, Denison MR. Expression and processing of nonstructural proteins of the human astroviruses. Adv Exp Med Biol. 1998;440:387–91.
25. Goodfellow I, Chaudhry Y, Gioldasi I, Gerondopoulos A, Natoni A, Labrie L, et al. Calicivirus translation initiation requires an interaction between VPg and eIF4. EMBO Rep. 2005;6(10): 968–72.
26. Gough RE, Borland ED, Keymer IF, Stuart JC. An outbreak of duck hepatitis type II in commercial ducks. Avian Pathol. 1985;14(2):227–36.
27. Gray EW, Angus KW, Snodgrass DR. Ultrastructure of the small intestine in astrovirus-infected lambs. J Gen Virol. 1980;49(1):71–82.
28. Guix S, Bosch A, Ribes E, Martínez LD, Pintó RM. Apoptosis in astrovirus-infected CaCo-2 cells. Virology. 2004;319(2):249–61.
29. Guix S, Caballero S, Bosch A, Pinto RM. C-Terminal nsP1a protein of human astrovirus colocalizes with the endoplasmic reticulum and viral RNA. J Virol. 2004;78(24):13627–36.
30. Guix S, Caballero S, Bosch A, Pintó RM. Human astrovirus C-terminal nsP1a protein is involved in RNA replication. Virology. 2005;333(1):124–31.
31. Herbert TP, Brierley I, Brown TD. Identification of a protein linked to the genomic and subgenomic mRNAs of feline calicivirus and its role in translation. J Gen Virol. 1997;78(Pt 5): 1033–40.
32. Jackson WT, Giddings TH, Taylor MP, Mulinyawe S, Rabinovitch M, Kopito RR, et al. Subversion of cellular autophagosomal machinery by RNA viruses. PLoS Biol. 2005;3(5):e156.
33. Jang SY, Jeong WH, Kim MS, Lee YM, Lee JI, Lee GC, et al. Detection of replicating negative-sense RNAs in CaCo-2 cells infected with human astrovirus. Arch Virol. 2010;155(9): 1383–9.
34. Jiang B, Monroe SS, Koonin EV, Stine SE, Glass RI. RNA sequence of astrovirus: distinctive genomic organization and a putative retrovirus-like ribosomal frameshifting signal that directs the viral replicase synthesis. Proc Natl Acad Sci USA. 1993;90(22):10539–43.
35. Jonassen CM, Jonassen TO, Saif YM, Snodgrass DR, Ushijima H, Shimizu M, et al. Comparison of capsid sequences from human and animal astroviruses. J Gen Virol. 2001;82(Pt 5):1061–7.
36. Jonassen CM, Jonassen TO, Sveen TM, Grinde B. Complete genomic sequences of astroviruses from sheep and turkey: comparison with related viruses. Virus Res. 2003;91(2):195–201.
37. Kiang D, Matsui SM. Proteolytic processing of a human astrovirus nonstructural protein. J Gen Virol. 2002;83(Pt 1):25–34.
38. Koci MD, Seal BS, Schultz-Cherry S. Molecular characterization of an avian astrovirus. J Virol. 2000;74(13):6173–7.
39. Koonin EV. The phylogeny of RNA-dependent RNA polymerases of positive-strand RNA viruses. J Gen Virol. 1991;72(Pt 9):2197–206.
40. Krishna NK. Identification of structural domains involved in astrovirus capsid biology. Viral Immunol. 2005;18(1):17–26.
41. Kuranaga E, Miura M. Nonapoptotic functions of caspases: caspases as regulatory molecules for immunity and cell-fate determination. Trends Cell Biol. 2007;17(3):135–44.

42. Lee TW, Kurtz JB. Serial propagation of astrovirus in tissue culture with the aid of trypsin. J Gen Virol. 1981;57(Pt 2):421–4.
43. Lewis TL, Greenberg HB, Herrmann JE, Smith LS, Matsui SM. Analysis of astrovirus serotype 1 RNA, identification of the viral RNA-dependent RNA polymerase motif, and expression of a viral structural protein. J Virol. 1994;68(1):77–83.
44. Lewis TL, Matsui SM. Astrovirus ribosomal frameshifting in an infection-transfection transient expression system. J Virol. 1996;70(5):2869–75.
45. Liu Y, Wimmer E, Paul A. Cis-acting RNA elements in human and animal plus-strand RNA viruses. Biochim Biophys Acta. 2010;1789:495–517.
46. Marczinke B, Bloys J, Brown TD, Willcocks MM, Carter MJ, Brierley I. The human astrovirus RNA-dependent RNA polymerase coding region is expressed by ribosomal frameshifting. J Virol. 1994;68(9):5588–95.
47. Miller S, Krijnse-Locker J. Modification of intracellular membrane structures for virus replication. Nat Rev Microbiol. 2008;6(5):363–74.
48. Monceyron C, Grinde B, Jonassen TO. Molecular characterisation of the 3′-end of the astrovirus genome. Arch Virol. 1997;142(4):699–706.
49. Monroe SS, Stine SE, Gorelkin L, Herrmann JE, Blacklow NR, Glass RI. Temporal synthesis of proteins and RNAs during human astrovirus infection of cultured cells. J Virol. 1991;65(2): 641–8.
50. Monroe SS, Jiang B, Stine SE, Koopmans M, Glass RI. Subgenomic RNA sequence of human astrovirus supports classification of astroviridae as a new family of RNA viruses. J Virol. 1993; 67(6):3611–4.
51. Moser LA, Schultz-Cherry S. Suppression of astrovirus replication by an ERK1/2 inhibitor. J Virol. 2008;82(15):7475–82.
52. Moser LA, Carter M, Schultz-Cherry S. Astrovirus increases epithelial barrier permeability independently of viral replication. J Virol. 2007;81(21):11937–45.
53. Méndez E, Aguirre-Crespo G, Zavala G, Arias CF. Association of the astrovirus structural protein VP90 with membranes plays a role in virus morphogenesis. J Virol. 2007;81(19): 10649–58.
54. Méndez E, Arias CF. Astrovirus. In: David M. Knipe, Peter M. Howley, editors. Fields virology. 6th ed. Philadelphia: Lippincott Williams & Wilkins; 2012. (in press).
55. Méndez E, Fernández-Luna MT, López S, Méndez-Toss M, Arias CF. Proteolytic processing of a serotype 8 human astrovirus ORF2 polyprotein. J Virol. 2002;76(16):7996–8002.
56. Méndez E, Salas-Ocampo MP, Munguía E, Arias CF. Protein products of the open reading frames encoding nonstructural proteins of human astrovirus serotype 8. J Virol. 2003;77(21): 11378–84.
57. Méndez E, Salas-Ocampo MP, Arias CF. Caspases mediate processing of the capsid precursor and cell release of human astroviruses. J Virol. 2004;78(16):8601–8.
58. Novoa RR, Calderita G, Arranz R, Fontana J, Granzow H, Risco C. Virus factories: associations of cell organelles for viral replication and morphogenesis. Biol Cell. 2005;97(2):147–72.
59. Pogue GP, Huntley CC, Hall TC. Common replication strategies emerging from the study of diverse groups of positive-strand RNA viruses. Arch Virol Suppl. 1994;9:181–94.
60. Roberts LO, Jopling CL, Jackson RJ, Willis AE. Viral strategies to subvert the mammalian translation machinery. Prog Mol Biol Transl Sci. 2009;90:313–67.
61. Sanchez-Fauquier A, Carrascosa AL, Carrascosa JL, Otero A, Glassl RI, Lopez JA, et al. Characterization of a human astrovirus serotype 2 structural protein (VP26) that contains an epitope involved in virus neutralization. Virology. 1994;201:312–20.
62. Satoh S, Hirota M, Noguchi T, Hijikata M, Handa H, Shimotohno K. Cleavage of hepatitis C virus nonstructural protein 5A by a caspase-like protease(s) in mammalian cells. Virology. 2000;270(2):476–87.
63. Speroni S, Rohayem J, Nenci S, Bonivento D, Robel I, Barthel J, et al. Structural and biochemical analysis of human pathogenic astrovirus serine protease at 2.0 A resolution. J Mol Biol. 2009;387(5):1137–52.

64. Steil BP, Barton DJ. Cis-active RNA elements (CREs) and picornavirus RNA replication. Virus Res. 2009;139(2):240–52.
65. Strauss JH, Strauss EG. The alphaviruses: gene expression, replication, and evolution. Microbiol Rev. 1994;58(3):491–562.
66. Sztuba-Solińska J, Stollar V, Bujarski JJ. Subgenomic messenger RNAs: mastering regulation of (+)-strand RNA virus life cycle. Virology. 2011;412(2):245–55.
67. Tanner NK, Linder P. DExD/H box RNA helicases: from generic motors to specific dissociation functions. Mol Cell. 2001;8(2):251–62.
68. Taylor P, Koci MD, Schultz-cherry S, Avian S. Astroviruses avian astroviruses. Avian Pathol. 2002;31(3):213–27.
69. Thal MA, Wasik BR, Posto J, Hardy RW. Template requirements for recognition and copying by Sindbis virus RNA-dependent RNA polymerase. Virology. 2007;358(1):221–32.
70. Velázquez-Moctezuma R, Banos-Lara MDR, Acevedo Y, Méndez E. Alternative cell lines to improve the rescue of infectious human astrovirus from a cDNA clone. J Virol Methods. 2011;179:295–302.
71. Villordo SM, Gamarnik AV. Genome cyclization as strategy for flavivirus RNA replication. Virus Res. 2009;139(2):230–9.
72. Wang QH, Kakizawa J, Wen LY, Shimizu M, Nishio O, Fang ZY, et al. Genetic analysis of the capsid region of astroviruses. J Med Virol. 2001;64(3):245–55.
73. Willcocks MM, Ashton N, Kurtz JB, Cubitt WD, Carter MJ. Cell culture adaptation of astrovirus involves a deletion. J Virol. 1994;68(9):6057–8.
74. Willcocks MM, Boxall AS, Carter MJ. Processing and intracellular location of human astrovirus non-structural proteins. J Gen Virol. 1999;80(Pt 10):2607–11.
75. Willcocks MM, Brown TD, Madeley CR, Carter MJ. The complete sequence of a human astrovirus. J Gen Virol. 1994;75(Pt 7):1785–8.
76. Willcocks MM, Carter MJ. Identification and sequence determination of the capsid protein gene of human astrovirus serotype 1. FEMS Microbiol Lett. 1993;114(1):1–7.
77. Wurzer WJ, Planz O, Ehrhardt C, Giner M, Silberzahn T, Pleschka S, et al. Caspase 3 activation is essential for efficient influenza virus propagation. EMBO J. 2003;22(11):2717–28.
78. Zhirnov OP, Konakova TE, Garten W, Klenk H. Caspase-dependent N-terminal cleavage of influenza virus nucleocapsid protein in infected cells. J Virol. 1999;73(12):10158–63.

Chapter 3
Astrovirus Structure and Assembly

Rebecca M. Dubois, Kelly A. Dryden, Mark Yeager, and Yizhi J. Tao

Abstract Recent structural studies on the astrovirus virion and viral proteins have yielded exciting new insights into the molecular mechanisms of the astrovirus life cycle. The 25 Å-resolution cryo-electron microscopy (Cryo-EM) reconstructions of the astrovirus virion reveal a solid capsid shell studded with spikes. Proteolytic maturation of the virus particle results in capsid conformational changes, most prominently at the spikes. High-resolution crystal structures of the human and avian astrovirus capsid spike domains have shed light on potential host receptors and species specificity. Together, both the structural studies on the astrovirus virion and capsid spike domains have revealed similarities to hepatitis E virus, suggesting an evolutionary relationship. The only other structural information on astrovirus is from the high-resolution crystal structure of the protease that is involved in nonstructural polyprotein processing. Overall, these structural studies have led a better understanding of the astrovirus life cycle, including astrovirus assembly, virus release, maturation, receptor binding, antibody neutralization, and nonstructural polyprotein processing.

R.M. Dubois, Ph.D. (✉)
Department of Structural Biology, St. Jude Children's Research Hospital, 262 Danny Thomas Place, Memphis, TN 38105, USA
e-mail: rebecca.dubois@stjude.org

K.A. Dryden, Ph.D.
Department of Molecular Physiology and Biological Physics, University of Virginia, Snyder Building, Room 232, Charlottesville, VA 22903, USA
e-mail: kdryden@virginia.edu

M. Yeager, M.D., Ph.D.
Department of Molecular Physiology and Biological Physics, University of Virginia, 1340 Jefferson Park Avenue, Charlottesville, VA 22908, USA
e-mail: yeager@virginia.edu

Y.J. Tao, Ph.D.
Department of Biochemistry and Cell Biology, Rice University, 6100 Main Street, MS140, Houston, TX 77005, USA
e-mail: ytao@rice.edu

S. Schultz-Cherry (ed.), *Astrovirus Research: Essential Ideas, Everyday Impacts, Future Directions*, DOI 10.1007/978-1-4614-4735-1_3,

Introduction

Structural studies on the astrovirus virion and viral proteins are important to understand the molecular details of the astrovirus life cycle. Recent structural studies of the astrovirus virion using cryo-electron microscopy have yielded important insights into the proteolytic maturation and corresponding conformational changes in the astrovirus virion. Additionally, recent high-resolution X-ray crystallographic studies on human and avian astrovirus capsid spikes have revealed insights into receptor binding, host cell entry, and species specificity. Together, the studies on the astrovirus virion and capsid spike have uncovered striking structural similarities to hepatitis E virus, suggesting that these viruses have a closer phylogenetic relationship than previously recognized. Finally, the high-resolution X-ray crystallographic structure of the astrovirus protease revealed a number of distinctive features that provide insight into its role in nonstructural polyprotein processing events. Overall, astrovirus structural studies provide a foundation for future mutagenesis studies to probe protein functions and roles in the astrovirus life cycle and pathogenesis. Furthermore, these structural studies will have a large impact on the development of antivirals and vaccines targeting astrovirus.

Cryo-EM Structure of Immature and Mature Human Astrovirus

Astroviruses are assembled from the structural protein VP90, an 87–90 kDa capsid precursor protein, which is encoded by the 3′-terminal one-third of the genome (ORF-2). This capsid protein has a highly conserved N-terminal domain and hypervariable C-terminal domain and must undergo a cascade of cleavages to become infectious. The final size and number of structural proteins that form the capsid vary depending on the serotype (reviewed by Krishna [23]). VP90 undergoes an initial caspase cleavage at the C-terminus to generate VP70 (a 70–79 kDa protein), which is required for particle release from cells [28]. Viral infectivity is further dependent upon trypsin cleavage and the genesis, usually, of three structural proteins, VP34, VP27 (27–29 kDa), and VP25 (25–26 kDa) [4, 27, 33]. The current consensus is that the cleavage product VP34 contains the conserved, N-terminal domain, while VP27 and VP25 contain the variable domain, which mediate the attachment of virions to cell surface receptors and contain epitopes for neutralizing antibodies [31]. VP25 is generated upon additional trimming at the N-terminus of VP27.

Astroviruses were initially named for the distinctive star-like appearance of the viral surface that was seen in ~10% of the fecally shed viral particles evaluated by negative-stain electron microscopy (EM) [2, 11]. EM ultrastructural analysis of infectious HAstV generated in cell culture showed spherical particles with a surface studded with spikes and an external diameter of 41 nm [30]. The characteristic star-like surface appearance was inducible after brief alkaline (pH 10) treatment of the particles, which made them less stable, and suggested that particles with star-like morphology may represent an intermediate step toward viral disassembly. More recently, the structures of immature (VP70) and mature virions have been calculated

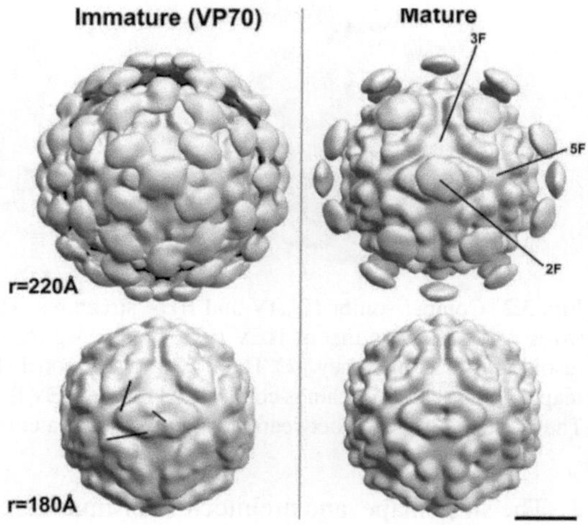

Fig. 3.1 Cryo-EM reconstructions of immature and mature HAstV. Immature (*left*) and mature (*right*) particles have very similar capsids as displayed by truncating the maps at radius 180 Å (*lower*), while the mature particles lack 60 of the 90 spikes visible on immature virions (*upper*). Symmetry axes are indicated on mature virion. One set of probable subunit contributors to the dimer spikes is displayed on the immature capsid. *Bar* = 100 Å

to moderate 25 Å-resolution by electron cryomicroscopy (cryo-EM) and image analysis [14]. The majority of the variable domain (a.a. 425–646) has been crystallized [13] and is discussed in detail later in this chapter.

Structural Characteristics and Similarity to Hepatitis E Virus

The cryo-EM reconstructions of immature assemblies of VP70 and mature, infectious particles both display two layers of density: an inner, continuous capsid layer and an outer layer composed of disconnected or weakly connected globular densities (Fig. 3.1). The capsids of the immature and mature particles are nearly identical at this resolution. The shells are ~350 Å in diameter and ~45 Å thick with features that support organization as $T=3$ icosahedral symmetry, therefore, assembled from 180 protein subunits. They have a solid, chiseled appearance forming triangular plateaus with the edges along the icosahedral twofold axes in close contact (Fig. 3.1, lower). There are depressions at the fivefold and threefold symmetry axes and a groove across the local twofold axes. The highly conserved N-terminal domain has a predicted secondary structure that would be homologous to the canonical β-sandwich fold in many other icosahedral viruses [23]. Therefore, the HAstV capsid shell is attributed to the conserved domain, which primarily comprises VP34 in the mature virions.

The distal layer of density is composed of discrete domains and represents the spikes that have been observed directly by electron microscopy (Fig. 3.1, upper). These globular domains have very poorly defined connections to the underlying capsid, implying that they may be flexible or have thin linkers. The densities are ~40 Å in diameter, although more elliptical than spherical, and extend to an outer radius of ~215 Å making the overall diameter of the particles ~430 Å. These densities are attributed the C-terminal, variable domain. Details of the astrovirus spike structure have been further elucidated from the crystal structure of the dimerized VP25 fragment [13].

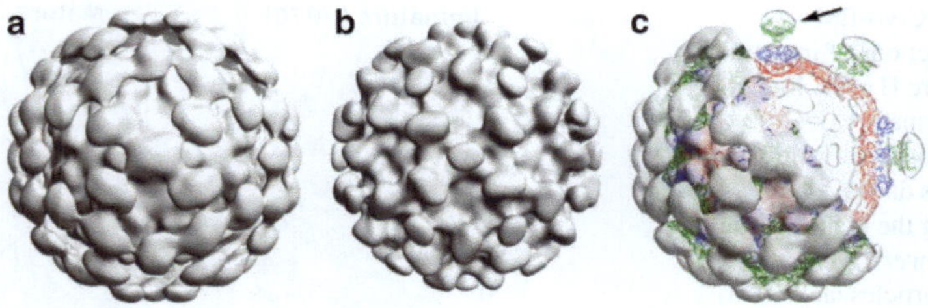

Fig. 3.2 Comparison of HAstV and HEV structures. The tertiary structure of immature HAstV (**a**) is very similar to that of HEV (**b**). The HEV-lp map (emd ID: 5,173) was truncated to 25 Å resolution for consistency. (**c**) The $T=3$ atomic model (PDB ID: 3IYO) docked into the HAstV map with no scaling. Chain is color coded for the HEV S- (*red*), M- (*blue*), and P-domains (*green*). The greatest difference between the maps is the greater radius of the HAstV spikes (*arrow*)

The size, shape, and architecture of immature HAstV is also remarkably similar to that of a $T=3$ hepatitis E virus-like particle (HEV-lp) calculated by cryo-EM [36] (Fig. 3.2a, b). Like astrovirus, HEV is a small, nonenveloped, single-stranded, positive-sense RNA virus with a genome size of about 7.2 kb. At present, it is the sole member of the *Hepeviridae* family. In addition to the cryo-EM reconstruction, crystal structures have also been determined for the HEV $T=1$ subviral particles assembled from recombinant proteins [19, 37]. The HEV structural protein has been described by sequential domains: a positively charged N-terminus likely involved in RNA packaging, a conserved β-sandwich forming the shell (S-domain), two protruding densities (P1- and P2-domains), and a C-terminal region that must be removed for in vitro assembly (the P1- and P2-domains are also referred to as the M- and P-domains, respectively). The $T=1$ crystal structure of HEV allowed detailed modeling of the protein subunits into the $T=3$ cryo-EM map. It was proposed that the capsid is assembled from dimers around the fivefold axes and across the icosahedral twofold axes. HEV using dimers as building blocks is further supported by the requirement for dimerization of the P2-domain for both assembly and host cell interactions [25, 37].

Like HEV, the overall structural organization of HAstV polyprotein can be similarly described, with the astrovirus conserved domain corresponding to the arginine-rich and S-domain of HEV, and the variable domain corresponding to the P1-, P2-, and cleaved C-terminal domains of HEV. Sequence alignment of the two viruses has 21, 13, and 14% identity for the S-, P1-, and P2-domains, respectively [14]. The major noticeable difference between the HAstV and HEV structures is that the astrovirus spikes extend to a greater radius than those of HEV, which has more compact spikes and an outer radius of ~190 Å (Fig. 3.2c, arrow). The HEV atomic model docks exceedingly well into the HAstV cryo-EM density maps without any scaling (Fig. 3.2c). The subunit organization of astrovirus can be inferred by analogy, and Dong et al. used the HEV atomic model to build a corresponding model for HAstV [13]. As with HEV, the spike stoichiometry and orientation suggests that they are dimers with contributions from subunits situated on different trimeric facets (Fig. 3.1, lower left). Using the nomenclature of HEV, AB dimers encircle the fivefold axes, while neighboring C subunits form dimers across the icosahedral twofold axes. In the case of HEV, and perhaps more so for HAstV, this arrangement leads to different interactions for the peripentonal dimers versus the dimers on the twofold axes. The spike bases for the twofold dimers are in much closer proximity than those around the fivefold, which are situated at nearly twice the distance across the local twofold groove. Extrapolating further from the HEV structure, it is likely that there is an underlying difference in the hinge angles between these dimers, which may reflect the presence or absence of RNA in close contact on the interior surface [36].

Partial Loss of Visible Spikes upon Trypsin Cleavage

While the capsid shells of both the immature and mature HAstV particles are essentially the same, there is a dramatic loss of spikes during trypsin cleavage (Fig. 3.1, upper). The reconstruction of the immature, uncleaved particles has 90 globular, or spike, densities. In the reconstruction of the mature particles, however, only the 30 spike densities located at the icosahedral twofold axes remain visible. Trypsin cleavage of the immature particles produces three products: we presume that VP34 forms the capsid shell and the two different C-terminal products, VP27 and VP25, form the surface spikes. Although a first assumption would be that the trypsin cleavage has removed two-thirds of the spikes from the particles, previously published SDS-PAGE gels show that the amount of VP34 present in infectious particles was comparable to that of VP27 and VP25 [4, 27, 33], indicating that most of the VP27 and VP25, although covalently independent, remain associated with the particle. Therefore, an alternative explanation is that the stoichiometry of the spikes may be conserved, and trypsin cleavage results in conformational flexibility so that the density is lost upon image averaging.

Proteolytic cleavage is a common mechanism for activation in both enveloped and nonenveloped viruses, including influenza virus [21, 24], coronavirus [15], rotavirus [16], reovirus [7], and alphaviruses [32]. A recurrent theme is that cleavage triggers a conformational change in a surface spike or other cell-attachment domain. In the case of rotavirus, the spike is initially completely unstructured, and therefore not observed by image analysis, until trypsin cleavage, at which point the polypeptides form a stable structure [12, 34]. It is possible that for astroviruses, the reverse takes place, and that fairly stable densities visible before cleavage become completely disordered at some positions. When the virus moves into the stomach of its host and is exposed to protease, the two arrangements of the spikes formed by AB and CC capsid protein dimers are likely to present different cleavage sites, giving rise to the two different C-terminal cleavage products, VP27 and VP25. The peripentonal spikes have dimer contributions that must extend nearly twice the distance of those at the icosahedral twofold axes, which may uncover sites for the additional trimming that generate VP25, and VP27 would be comprise the remaining visible spikes. The spikes located on the icosahedral twofolds are likely to be

VP27 homodimers, which have restricted movement just like the spikes in unprocessed capsid and thus can be visualized in a population-averaged reconstruction. There are several potential ways the conformational change could confer infectivity—by exposing the capsid surface, allowing some variable domains to be flexible, or exposure of binding sites.

Crystal Structure of the Human Astrovirus Capsid Spike

Structure Overview

Among the four small, nonenveloped positive-sense RNA viruses (i.e., picornavirus, calicivirus, astrovirus, and hepevirus) that infect animals, astrovirus is the only one whose atomic structure is not yet fully resolved. Based on primary sequence analysis, it was hypothesized that VP34 builds up the continuous capsid shell, while VP27 and VP25 form the dimeric projections seen in the cryo-EM reconstruction image [14, 23]. In an attempt to express the projection/spike domain, Dong et al. made five constructs containing residues 394–646, 501–658, 415–628, 501–628, and 415–646, respectively [13]. Of these five constructs, only $P2^{415-646}$ produced soluble proteins in *E. coli*, presumably because it encompasses an intact spike domain, and its two ends both terminate in flexible domain linkers. Considering that the N-terminal ends of VP27 and VP25 start from Gln394 and Ser424, respectively [27], $P2^{415-646}$ is a little longer than VP25 but a bit shorter than VP27 at the N-terminus. $P2^{415-646}$ exists as a dimer in solution as judged by gel filtration chromatography [13].

Purified $P2^{415-646}$ dimer was crystallized in two different crystal forms of space group $P3_221$ and $I2_13$, respectively. The $I2_13$ crystals, grown at pH 4.6, diffracted to only 2.8 Å possibly due to its high solvent content (82.8%). The $P3_221$ crystals, which were grown at pH 9.5, diffracted far better to 1.8 Å resolution (solvent content = 41.0%). Both structures have been determined, but it was found that the two structures were essentially the same with a RMSD of only 0.548 Å for all Ca atoms, despite the fact that these two crystals were grown at very different pHs and had different crystal packing interactions. Amino acid residues 430–645 of $P2^{415-646}$ were modeled in the crystal structure. As in solution, $P2^{415-646}$ formed dimers in both crystal forms around the twofold crystal symmetry axes. As discussed below, the crystal structure of the astrovirus $P2^{415-646}$ spike provides a first glimpse into the astrovirus capsid structure at the atomic level (PDB ID: 3QSQ).

The astrovirus surface spike has an exclusively β-stranded structure. Each $P2^{415-646}$ molecule contains a core, six-stranded β-barrel (β1, β4, β6, β7, β10, and β11) with a tightly packed hydrophobic core (Fig. 3.3). Another three-stranded β-sheet, consisting of β2, β3, and β5, is packed against the outside of the β-barrel away from the dimer interface. Therefore, the structure of $P2^{415-646}$ can be formalized as a triple-layered β-sandwich, in which the three orthogonal β-sheets are connected by two highly bent β-strands (β4 and β6). Two long loops, L1 and L4 (L1: residues

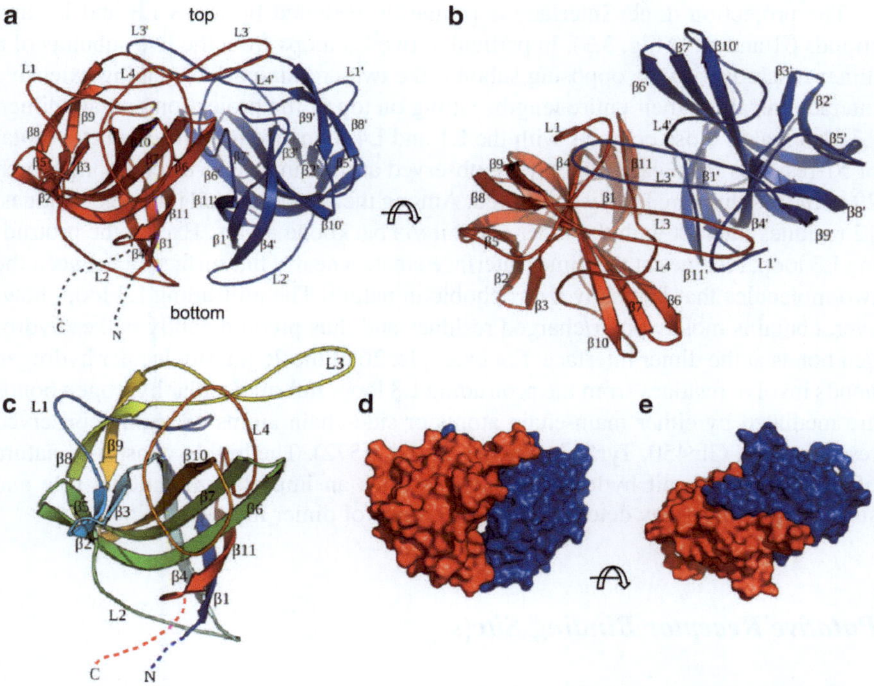

Fig. 3.3 Crystal structure of astrovirus P2$^{415–646}$. (**a**) Side view of the P2$^{415–646}$ dimer. The two subunits are colored in *red* and *blue*, respectively. The N- and C-termini of the *red* subunit are highlighted by *dotted lines*. The viral capsid lies at the bottom the dimeric spike. (**b**) Top view of the P2$^{415–646}$ dimer. (**c**) Structure of one P2$^{415–646}$ molecule. The polypeptide is *rainbow*-colored with the N-terminus in *blue* and the C-terminus in *red*. (**d**, **e**) Surface representations of the P2$^{415–646}$ dimer for comparison with the cryo-EM reconstruction image. Molecules in (**d**) and (**e**) are viewed from the same orientation as in (**a**) and (**b**), respectively

439–466; L4: residues 597–632), span across the top of the β barrel, while a shorter loop, L2 (residues 491–506), runs across the bottom of the barrel (Fig. 3.3). A β-hairpin motif, consisting of β8, β9, and L3 (residues 554–574), sits on top of L1 and L4, and may help to reinforce the structural conformation of these two long loops.

The P2$^{415–646}$ dimer appears to be very stable. The buried surface area at the dimer interface is ~3,490 Å^2, which counts for 20.2% of the solvent-accessible area of the complex. Therefore, the dimer interaction mediated by the projection domain is likely to play an important role in stabilizing the viral capsid. Because the astrovirus projection domain is connected to the capsid shell at its N-terminus, the thinner end of the projection dimer (where both the N- and C-termini are located) should directly abut the capsid outer surface, whereas the opposite, flat end would point away from the capsid (Fig. 3.3a). When visually compared to a cryo-EM reconstruction image of an astrovirus (Fig. 3.1), the overall morphology of the P2$^{415–646}$ dimer matches that of the dimeric surface spikes.

The projection dimer interface is primarily mediated by loops L3 and L4 and strands β1 and β11 (Fig. 3.3). In particular, two L3 loops from the two subunits of a dimer reach over to the opposing subunit like two crossing arms, making extensive interactions along their entire length. Sitting on top of the projection domain dimer, L3 also makes close contacts with the L1 and L4 loops from both subunits. A total of 51 residues from each subunit are observed at the dimer interface, among which 29 of the residues are highly conserved. Among the nonconserved interface residues, 12 residues interact with the other subunit via backbone atoms. Except the protruding L3 loop, residues at the dimer interface create a nearly flat surface in between the two molecules that is mostly hydrophobic in nature. The protruding L3 loop, however, contains mostly polar/charged residues and thus predominantly makes hydrogen bonds at the dimer interface. For example, 20 of the 26 intermolecular hydrogen bonds involve residues from the protruding L3 loop. Intermolecular hydrogen bonds are mediated by either main-chain atoms or side-chain atoms of highly conserved residues (i.e., Gln450, Tyr555, Arg558, and Arg572). The highly conserved nature of these inter-subunit hydrogen bonds indicates an important functional role and suggests that they may determine the specificity of dimer interaction.

Putative Receptor Binding Site(s)

Little is known about astrovirus receptor(s) and virus-receptor binding, although previous antigenicity and sequence analysis implicate the P2 domain in cell attachment and receptor binding [5, 23, 33]. Aligning the human astrovirus $P2^{415-646}$ domain of serotype 8 to that of the other seven serotypes gave pairwise percentage identities ranging from 47% to 57%. It is likely that all human astroviruses use a common receptor for cell internalization, even different astrovirus serotypes might use different cell surface molecules for initial attachment [9]. By calculating the amino acid conservation scores for the P2 domain using the sequences of all eight human astrovirus serotypes, several conserved regions were observed [13]. The most interesting one, which was called the P-site, is located in a shallow groove on top of the P2 spike (Fig. 3.4). This region has a large number of conserved residues (Lys455, Ser 554, Thr575, and Glu610 strictly conserved in all eight serotypes), all with their side chains exposed to solvent. The location of this site also makes it highly accessible to cell receptors without steric hindrance. Another conserved surface patch, the S-site, is located on the side of the P2 spike. This site contains three strictly conserved aromatic residues (Tyr475, Trp604, and Trp606), but two of these (Trp604 and Trp606) have their side chains tucked inside and thus may have been retained for the folding of the protein rather than for ligand recognition. Interestingly, an exposed β-turn (Thr529–Asn530–Asn531–Arg532) between strands β6 and β7 is also well conserved. This β-turn and its adjacent regions have the highest B-factors among the entire structure.

A close inspection of the P-site shows that it is largely hydrophilic in nature, and is abundant in hydrogen bond donors and acceptors (Fig. 3.4d). In addition to the strictly conserved residues Lys455, Ser 554, Thr575, and Glu610, other residues

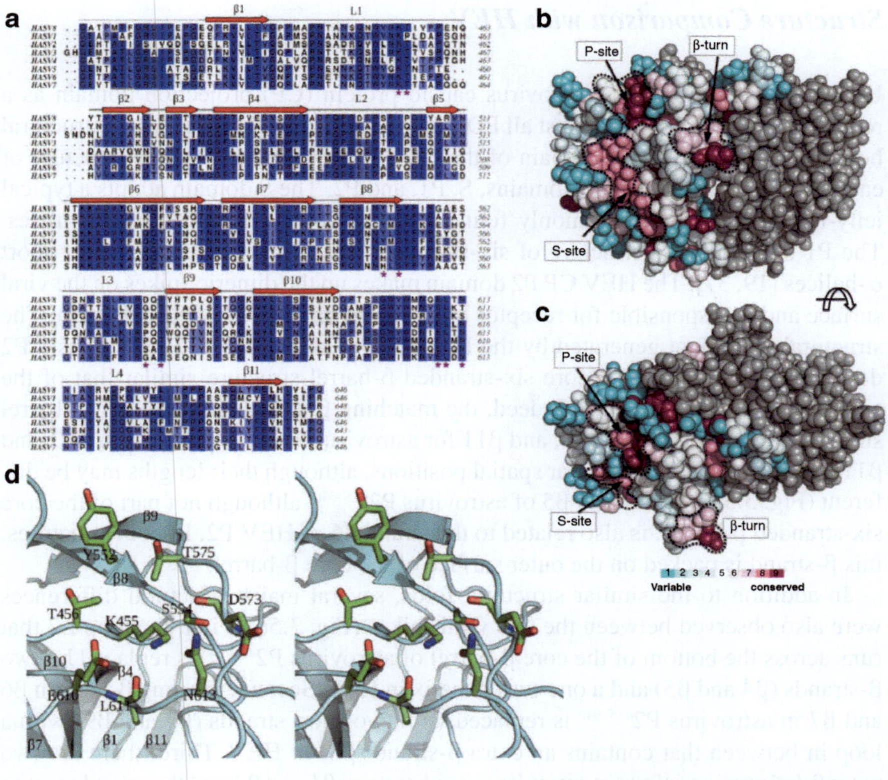

Fig. 3.4 Conserved surfaces on the human astrovirus projection domain. (**a**) Multiple sequence alignment of the projection domains (residues 415–646 for serotype 8) from human astroviruses of the eight different serotypes. The level of conservation is color-coded by different shades of *white* (least conserved) and *blue* (strictly conserved). The multiple sequence alignment was computed using Aline [8]. (**b, c**) Astrovirus projection domain, viewed from the side and the top, respectively. Residues in one subunit are colored according to their levels of conservation. The other subunit is colored in *gray*. Three highly conserved sites are circled by *dotted black curves*. (**d**) Stereo view of the putative receptor-binding site (P-site)

found in the vicinity are Thr456 (Pro in serotype 1), Tyr552 (His in serotypes 5, 6, and 7), and Asn613 (Thr in serotype 1, Ser in serotype 2), together with main-chain carbonyls from Asp573 and Leu611. These residues mainly come from the three loops L1, L3, and L4 (Fig. 3.4d). The polar residue composition, especially the positively/negatively charged pair Lys455 and Glu610, is also observed in the receptor binding sites of several viruses that use oligosaccharide moieties as receptors (e.g., influenza A virus and norovirus) [10, 17], suggesting that human astroviruses are likely to use a polysaccharide molecule as a cell receptor. Further molecular modeling and docking indicate that the P-site may be able to accommodate a di- or trisaccharide moiety. Consistent with structural analysis, the addition of heparin, heparan sulfate, or dextran sulfate was able to partially block HAstV-8 infectivity in a consistent manner [13]. Among the three polysaccharides used, heparin had the most pronounced effect.

Structure Comparison with HEV

Using the structure of the astrovirus capsid protein (CP) projection domain as a reference, a DALI search against all PDB entries indicated that the closest structural homolog is the projection domain of the HEV CP (z-score = 3.9). The structure of each HEV CP contains three domains, S, P1, and P2. The S domain adopts a typical jelly-roll β-barrel fold commonly found in other positive-stranded RNA viruses. The P1 domain has a structure of six-stranded β-barrel decorated with four short α-helices [19, 37]. The HEV CP P2 domain makes up the dimeric spikes on the viral surface and is responsible for receptor binding and antigenicity determination. The structural alignment generated by the DALI server shows that the HEV capsid P2 domain [19, 25, 37] has a core six-stranded β-barrel structure similar that of the astrovirus $P2^{415-646}$ (Fig. 3.5). Indeed, the matching β-strands in these two β-barrel structures (β1, β4, β6, β7, β10, and β11 for astrovirus, and β2, β3, β7, β9, β10, and β11 for HEV) have very similar spatial positions, although their lengths may be different (Fig. 3.5b). The strand β5 of astrovirus $P2^{415-646}$, although not part of the core six-stranded β-barrel, is also related to the strand β6 of HEV P2. In both structures, this β-strand is packed on the outer surface of the core β-barrel.

In addition to the similar structural folds, several major structural differences were also observed between the two viral spikes (Fig. 3.5b). First, the loop L2 that runs across the bottom of the core β-barrel of astrovirus $P2^{415-646}$ is replaced by two β-strands (β4 and β5) and a one-turn α-helix in HEV. Second, the long β-hairpin β6 and β7 in astrovirus $P2^{415-646}$ is replaced with two short strands (β7 and β9) with a loop in between that contains an extra β-strand (β8) in HEV. Third, there are two extra β-hairpin insertions in astrovirus, one between β1 and β4 and the other between β7 and β10. Fourth, β6, β7, and β10 in astrovirus CP are much longer than their counterparts in HEV P2. In astrovirus, these three β-strands form a small protruding knob on each side of the surface spikes. Because of these insertions and also because of its longer β-strands, the astrovirus P2 domain is substantially larger in size compared to the HEV P2 (232 vs. 143 residues), which could account for the extended radius of the HAstV spikes (Fig. 3.2).

The P2 dimer interaction is mediated by similar structural elements in both astrovirus and HEV, as structure superposition based on P2 monomers also placed the P2 dimers on top of each other, with the two dimer axes differing by only ~14°. Therefore, the inclination angle of the β-barrel with respect to the icosahedral twofold symmetry axis must be similar in astrovirus and HEV, but the azimuthal rotation of the dimeric spike about the icosahedral twofold axis is rather different in these two viruses. Like in the HEV CP, both N- and C-termini of the astrovirus P2 are located at the bottom of the spike close to the particle surface.

The detailed structural similarity between the astrovirus and HEV surface spikes suggests that the two viruses may have a much closer phylogenetic relationship than previously expected [22]. On the other hand, the caliciviruses, another group of small, nonenveloped, positive-sense RNA viruses, also have dimeric surface spikes. According to Dali, calicivirus CP is also related to astrovirus CP in its overall

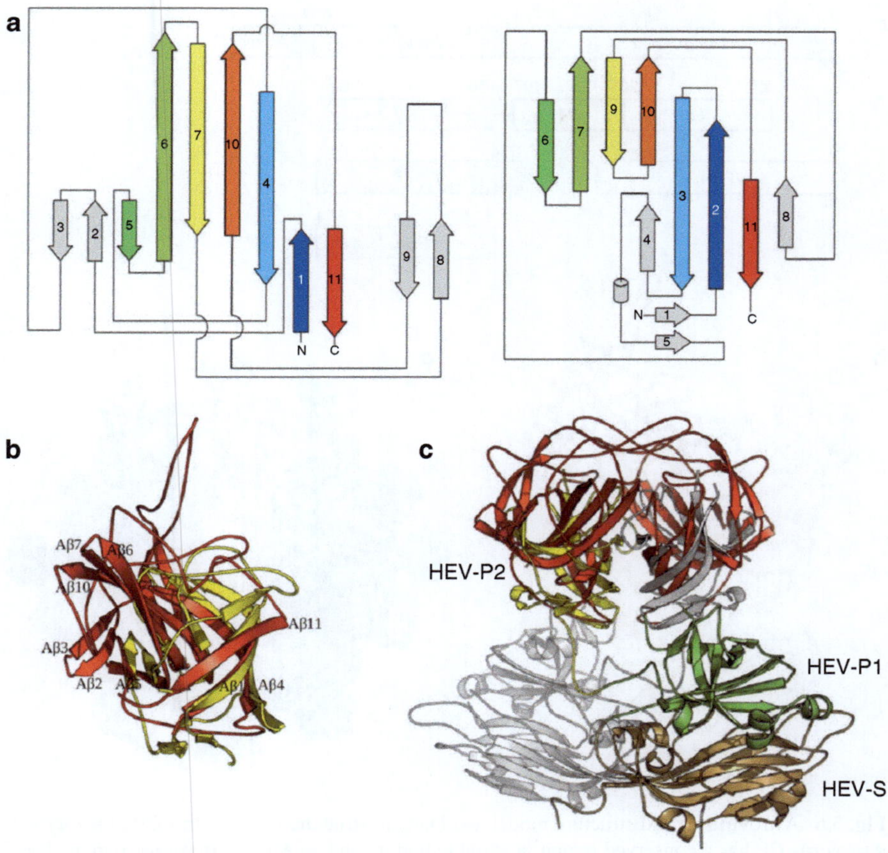

Fig. 3.5 Structural comparison of P2$^{415-646}$ with HEV capsid P2 domain. (**a**) Topology diagrams for the astrovirus P2$^{415-646}$ (*left*) and the HEV P2 domain (*right*). Related β-strands in the two structures are rendered in matching colors, while unrelated structural elements are shaded in *gray*. (**b**) Superposition of the astrovirus P2$^{415-646}$ (*red*) and the HEV P2 domain (*yellow*). Structure elements in astrovirus P2$^{415-646}$ are labeled as in Fig. 3.3c. with a letter "A". (**c**) Superposition of the astrovirus P2$^{415-646}$ dimer with the HEV CP dimer. The astrovirus P2$^{415-646}$ dimer is in *red*. For HEV CP, one subunit is in *gray*, and the other is colored by domains: S in *gold*; P1 in *green*; and P2 in *yellow*

structural fold, but the level of similarity is significantly less than that for HEV ($Z=1.7$ vs. $Z=3.9$). In addition, the domain organization of the calicivirus CP is different compared to that of astrovirus and HEV CP. While the three domains of the HEV CP are organized in a linear fashion, the polypeptide of the calicivirus CP winds through the three domains in the order of S → P1 → P2 → P1 [29]. Therefore, the surface spike of calicivirus is formed effectively by a polypeptide insertion in the P1 domain. In contrast, the spike of both HEV and astrovirus is situated on the CP polypeptide after the P1 domain (Fig. 3.6a). Based on these observations, astrovirus is likely to have more distant phylogenetic relationship to calicivirus compared to HEV, at least when the viral CPs are used for analysis.

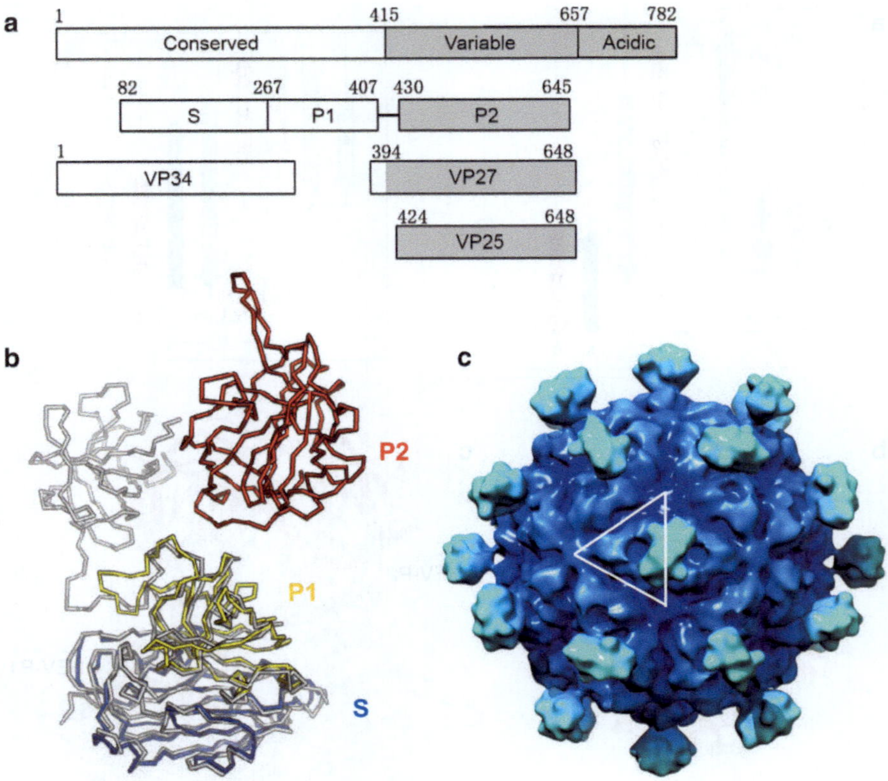

Fig. 3.6 Astrovirus capsid structure model. (**a**) Domain structure of astrovirus CP of serotype 8. Astrovirus CP has a conserved region, a variable region, and an acidic C-terminal region. The S and P1 domain of astrovirus can be roughly mapped to the conserved region. The P2 domain, however, is mapped to the variable region. The coverage of VP34, VP27, and VP24 is also indicated. (**b**) Astrovirus CP superimposed onto HEV CP. HEV CP is shown in *gray*, whereas the astrovirus CP is colored by domains with S in *blue*, P1 in *yellow*, and P2 in *red*. (**c**) $T = 3$ astrovirus capsid model rendered at 15 Å resolution. The *white lines* highlight one asymmetric unit

Structure Modeling of the Human Astrovirus Capsid and Insights into Maturation

Based on the noted structural similarity between the astrovirus and HEV CPs, an attempt was made to create an atomic structure of the astrovirus CP by homology modeling [13]. Indeed, homology structure recognition based on the astrovirus CP sequence alone indicated that the HEV CP is the best structural homolog [13]. In particular, residues 82–266 and 267–407 of the human astrovirus (serotype 8) CP were matched to the S (aa118–313) and P1 (aa314–453) domains of the HEV CP, respectively. Based on this alignment, a homology model of the human astrovirus CP was created with the HEV CP structure as a template [13] (Fig. 3.6). By aligning the S and P1 domains of the astrovirus CP model onto the three quasi-equivalent

HEV CP molecules from the $T=3$ EM model [36], a $T=3$ astrovirus capsid structure was generated [13]. The astrovirus capsid model closely resembles the cryo-EM reconstruction of astrovirus in terms of particle diameter and the capsid landscape variation. The close resemblance of the two structures further confirms the notion that astroviruses are phylogenetically related to HEV.

During maturation, the immature viral particles are processed by extracellular proteases to produce smaller peptides species that are observed in infectious particles. VP34 is derived from the conserved region, while VP27 and VP25 are from the hypervariable region of the astrovirus CP [26] (Fig. 3.6). The N-terminal end of VP27 and VP25 starts from Gln394 and Ser424, respectively [27]. Based on the astrovirus capsid model, the N-terminal cleavage site of VP27 (Gln394) was mapped to the unstructured loop right before the last β-strand in the P1 domain. Therefore, VP27 should be able to remain associated with the viral particle through main-chain H-bonds mediated by this last β-strand in the P1 domain. The N-terminal cleavage site of VP25, however, is located in an extended linker (the P1–P2 linker) between the P1 and the P2 domain. Therefore, VP25 may need to dimerize with a VP27 to remain stably bound to the viral particle.

While the astrovirus CP is larger than HEV CP by ~120aa, there may be additional similarities in their capsid assembly processes. HastV-8 has been shown to have caspase processing at the $TYVD_{657}$ sequence [3], which removes the last ~125 amino acids for the VP90 conversion to VP70 and is required for viral release. Although this particular sequence is not conserved in all astroviruses, they all contain potential caspase cleavage sequences within amino acids 650–700 of their CP polypeptides. Because HEV assembly has only been achieved in vitro with the deletion of ~50 C-terminal amino acids, HEV may have a similar cleavage requirement for particle release. As very little is known about HEV infectivity, it is tempting to speculate that those viruses may also require extracellular protease cleavage for activation, which may be similarly asymmetric to that observed for astrovirus.

Avian Astrovirus Capsid Spike Domain Structure

Recently, the crystal structure of the turkey astrovirus 2 capsid spike domain was determined at 1.5 Å resolution and publicly released by the Protein Data Bank (PDB ID: 3TS3). Although the corresponding publication is not yet published, visual inspection of the structure reveals significant structural divergence compared to the HAstV-8 capsid spike structure (PDB ID: 3QSQ) [13]. Whereas the human spike domain is composed entirely of β-strands and forms a bowl-shaped dimeric spike, the avian spike domain contains several helices as well as β-strands and forms a heart-shaped dimeric spike (Fig. 3.7). In fact, a structural homology search identifies the HEV spike domain, not the human astrovirus spike domain, as the closest structural homolog of the avian astrovirus spike domain. The low sequence homology and low structural conservation between the mammalian and avian astrovirus spike domains suggest that mammalian and avian astroviruses may have significant mechanistic differences in host cell entry.

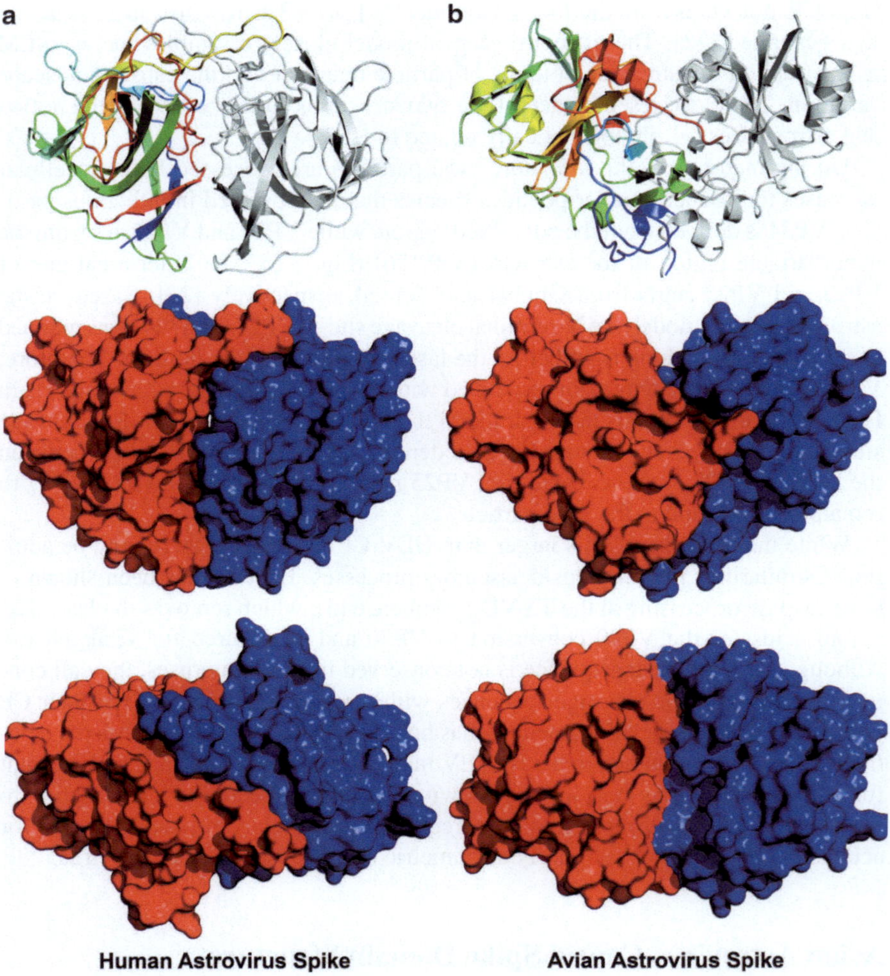

Fig. 3.7 Comparison of human and avian astrovirus capsid spike structures. (**a**) Dimeric structure of the human astrovirus 8 capsid spike shown as cartoon (*top*) or surface (*bottom*) representation. (PDB entry 3QSQ). (**b**) Dimeric structure of the turkey astrovirus 2 spike shown as cartoon (*top*) or surface (*bottom*) representation (PDB entry 3TS3). In the cartoon representations, one protomer is colored in a *blue*-to-*red rainbow* from N- to C-termini whereas the other protomer is colored *gray*. In the surface representations, one protomer is colored *red* and the other is colored *blue*

Human Astrovirus Protease Structure

The astrovirus genome open reading frames ORF1a and ORF1b encode the non-structural proteins nsp1a and nsp1b, respectively, which are involved in viral replication. The nonstructural proteins are synthesized as polyprotein precursors that become proteolytically processed. Although the exact processing sites remain

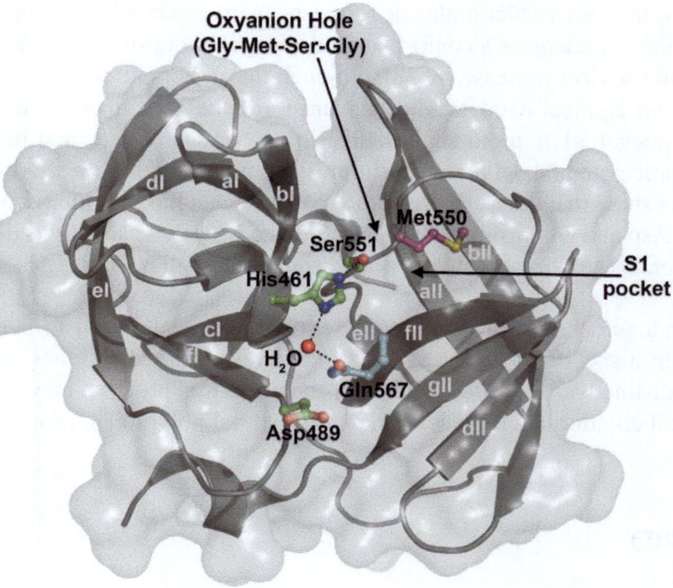

Fig. 3.8 Structure of the astrovirus protease. Human astrovirus 1 protease structure (ORF1a polyprotein residues 432–587) shown as cartoon representation overlaid with a semitransparent surface representation. The catalytic triad (His461–Asp489–Ser551) side chains are shown as *green ball*-and-stick models. The Gln567 side chain that hydrogen bonds to a water molecule in the active site is shown as *cyan ball*-and-*stick* model. The water molecule is shown as a *red sphere*. The Met550 side chain of the oxyanion hole (Gly549–Met550–Ser551–Gly 552) is shown as *magenta ball*-and-*stick* model. β-Strands are labeled in *light gray*. Locations of the oxyanion hole and positively charged S1 substrate binding pocket are described with *arrows*

undefined, a 3C-like serine protease motif was found to be encoded at the 3′-end of ORF1a and is thought to play an important role in the processing events [20]. Further studies show that mutation of the astrovirus protease motif resulted in deficient or altered processing of nsp1a translated in vitro [20] or in BHK cells [18].

The human astrovirus 1 protease crystal structure was determined at 2.0 Å resolution (PDB entry 2W5E) [35]. The 17 kDa protein is a member of the trypsin-like serine protease family and contains two β-barrel domains connected by a long interdomain loop (Fig. 3.8). The active site contains the catalytic triad His461, Asp489, and Ser551 that was predicted by Kiang et al. (1992), and these residues are strictly conserved in all astroviruses [20]. In addition, the Gly–X–Ser–Gly motif conserved in all astroviruses at residues 549–552 that forms the oxyanion hole is in the correct structural conformation for stabilization of the intermediate during catalysis.

However, the astrovirus protease has several structural features that distinguish it from other viral proteases [35]. First, two loops near the active site, loop eI-fI and loop cII-dII, are shorter than most viral proteases. The result of this is a more open and solvent-exposed active site. Second, the active site residue Asp489 is oriented away from His461 and does not form the hydrogen bond normally seen in trypsin-like proteases. Instead, a water molecule forms the hydrogen bond. The authors note that Asp489 likely undergoes a conformational change during substrate binding, as seen in hepatitis A virus protease [6, 38], although the authors do not rule out the possibility of an atypical catalytic mechanism. Last, the astrovirus protease substrate binding pocket S1 is more shallow than other viral proteases and has a positive electrostatic potential. Consistent with this structural observation, hydrolytic activity assays show that the astrovirus protease S1 pocket has specificity for negatively charged Asp and Glu residues. Furthermore, the hydrolytic activity is dependent on a functional astrovirus protease active site because the Ser551Ala mutant was inactive.

Taken together, the astrovirus protease structure reveals a trypsin-like serine protease with a specificity for cleavage at negatively charged residues Asp and Glu. Several distinctive structural features of the astrovirus protease may allow for further specificity during cleavage of the nonstructural polyprotein processing events.

Summary

Astrovirus structure and assembly are particularly intriguing in several aspects. First, a ~125aa C-terminal domain of the viral CP is removed by host caspase following assembly. The exact function of this C-terminal domain is not known but it may facilitate the sorting/trafficking of the viral CP to promote assembly. Additionally, extensive extracellular protease processing of the viral capsid is required for virus infectivity and maturation, which is rather unusual for small, positive-sense RNA viruses. In this chapter, we have primarily focused our discussion on the cryo-EM reconstructions of the immature and mature astrovirus, the crystal structures of the human astrovirus and turkey astrovirus capsid spikes, and the crystal structure of the human astrovirus protease. Altogether, these new findings have led a better understanding of many facets of the astrovirus replication, including astrovirus assembly, virus release, maturation, receptor binding, antibody neutralization, and nonstructural polyprotein processing. Of course, these findings would not be possible without many years of pioneering virological work that is not fully discussed here. We also realize that the ultimate understanding of the astrovirus assembly and maturation will require further characterization of the viral capsid/capsid proteins by structural as well as biochemical, genetic, and cell biology approaches.

Although there is a clear structural homology between the astrovirus and HEV capsid proteins, the replication mechanism of the astrovirus appears to differ from that of HEV. Unlike HEV, which is known to encode both methyltransferase and guanylyltransferase activities within the 5′ half of its ORF1, no capping enzymes have been identified in the astrovirus nonstructural proteins. Only three enzymes, the viral protease, a viral polymerase, and a putative helicase, have been found in the astrovirus genome. Interestingly, phylogenetic analysis suggests that the astrovirus genome might encode a VPg with sequence similar to the VPgs from *Caliciviridae* and *Potyviridae* [1]. We expect that further studies on astrovirus and other small, nonenveloped RNA viruses, both on their capsid and their replication, would help to establish their phylogenetic relationships, which, in turn, would greatly facilitate further structural and functional characterization of these viruses.

Acknowledgments The authors are supported by the Welch Foundation (C-1565 to YJT) and the National Institutes of Health (R01 GM066087 to MY).

References

1. Al-Mutairy B, Walter JE, Pothen A, Mitchell DK. Genome prediction of putative genome-linked viral protein (VPg) of astroviruses. Virus Genes. 2005;31:21–30.
2. Appleton H, Higgins PG. Letter: Viruses and gastroenteritis in infants. Lancet. 1975;1:1297.
3. Banos-Lara Mdel R, Méndez E. Role of individual caspases induced by astrovirus on the processing of its structural protein and its release from the cell through a non-lytic mechanism. Virology. 2010;401:322–32.
4. Bass DM, Qiu S. Proteolytic processing of the astrovirus capsid. J Virol. 2000;74:1810–4.
5. Bass DM, Upadhyayula U. Characterization of human serotype 1 astrovirus-neutralizing epitopes. J Virol. 1997;71:8666–71.
6. Bergmann EM, Mosimann SC, Chernaia MM, Malcolm BA, James MN. The refined crystal structure of the 3C gene product from hepatitis A virus: specific proteinase activity and RNA recognition. J Virol. 1997;71:2436–48.
7. Bodkin DK, Nibert ML, Fields BN. Proteolytic digestion of reovirus in the intestinal lumens of neonatal mice. J Virol. 1989;63:4676–81.
8. Bond CS, Schuttelkopf AW. ALINE: a WYSIWYG protein-sequence alignment editor for publication-quality alignments. Acta Crystallogr D Biol Crystallogr. 2009;65:510–2.
9. Brinker JP, Blacklow NR, Herrmann JE. Human astrovirus isolation and propagation in multiple cell lines. Arch Virol. 2000;145:1847–56.
10. Cao S, Lou Z, Tan M, Chen Y, Liu Y, Zhang Z, et al. Structural basis for the recognition of blood group trisaccharides by norovirus. J Virol. 2007;81:5949–57.
11. Caul EO, Appleton H. The electron microscopical and physical characteristics of small round human fecal viruses: an interim scheme for classification. J Med Virol. 1982;9:257–65.
12. Crawford SE, Mukherjee SK, Estes MK, Lawton JA, Shaw AL, Ramig RF, et al. Trypsin cleavage stabilizes the rotavirus VP4 spike. J Virol. 2001;75:6052–61.
13. Dong J, Dong L, Méndez E, Tao Y. Crystal structure of the human astrovirus capsid spike. Proc Natl Acad Sci USA. 2011;108:12681–6.
14. Dryden KA, Tihova M, Nowotny N, Matsui SM, Méndez E, Yeager M. Immature and mature human astrovirus: structure, conformational changes, and similarities to Hepatitis E Virus. J Mol Biol. 2012; Epub ahead of print June 25.
15. Du L, Kao RY, Zhou Y, He Y, Zhao G, Wong C, et al. Cleavage of spike protein of SARS coronavirus by protease factor Xa is associated with viral infectivity. Biochem Biophys Res Commun. 2007;359:174–9.
16. Estes MK, Graham DY, Mason BB. Proteolytic enhancement of rotavirus infectivity: molecular mechanisms. J Virol. 1981;39:879–88.
17. Gamblin SJ, Haire LF, Russell RJ, Stevens DJ, Xiao B, Ha Y, et al. The structure and receptor binding properties of the 1918 influenza hemagglutinin. Science. 2004;303:1838–42.
18. Geigenmüller U, Ginzton NH, Matsui SM. Studies on intracellular processing of the capsid protein of human astrovirus serotype 1 in infected cells. J Gen Virol. 2002;83:1691–5.
19. Guu TSY, Liu Z, Ye Q, Mata DA, Li K, Yin C, et al. Structure of the hepatitis E virus-like particle suggests mechanisms for virus assembly and receptor binding. Proc Natl Acad Sci USA. 2009;106:12992–7.

20. Kiang D, Matsui SM. Proteolytic processing of a human astrovirus nonstructural protein. J Gen Virol. 2002;83:25–34.
21. Klenk HD, Rott R, Orlich M, Blodorn J. Activation of influenza A viruses by trypsin treatment. Virology. 1975;68:426–39.
22. Koci MD, Seal BS, Schultz-Cherry S. Molecular characterization of an avian astrovirus. J Virol. 2000;74:6173–7.
23. Krishna NK. Identification of structural domains involved in astrovirus capsid biology. Viral Immunol. 2005;18:17–26.
24. Lazarowitz SG, Choppin PW. Enhancement of the infectivity of influenza A and B viruses by proteolytic cleavage of the hemagglutinin polypeptide. Virology. 1975;68:440–54.
25. Li S, Tang X, Seetharaman J, Yang C, Gu Y, Zhang J, et al. Dimerization of hepatitis E virus capsid protein E2s domain is essential for virus-host interaction. PLoS Pathog. 2009;5:e1000537.
26. Méndez E, Aguirre-Crespo G, Zavala G, Arias CF. Association of the astrovirus structural protein VP90 with membranes plays a role in virus morphogenesis. J Virol. 2007;81:10649–58.
27. Méndez E, Fernández-Luna T, López S, Méndez-Toss M, Arias CF. Proteolytic processing of a serotype 8 human astrovirus ORF2 polyprotein. J Virol. 2002;76:7996–8002.
28. Méndez E, Salas-Ocampo E, Arias CF. Caspases mediate processing of the capsid precursor and cell release of human astroviruses. J Virol. 2004;78:8601–8.
29. Prasad BV, Hardy ME, Dokland T, Bella J, Rossmann MG, Estes MK. X-ray crystallographic structure of the Norwalk virus capsid. Science. 1999;286:287–90.
30. Risco C, Carrascosa JL, Pedregosa AM, Humphrey CD, Sánchez-Fauquier A. Ultrastructure of human astrovirus serotype 2. J Gen Virol. 1995;76(Pt 8):2075–80.
31. Royuela E, Sánchez-Fauquier A. Molecular cloning, expression and first antigenic characterization of human astrovirus VP26 structural protein and a C-terminal deleted form. Comp Immunol Microbiol Infect Dis. 2010;33:1–14.
32. Salminen A, Wahlberg JM, Lobigs M, Liljestrom P, Garoff H. Membrane fusion process of Semliki Forest virus. II: cleavage-dependent reorganization of the spike protein complex controls virus entry. J Cell Biol. 1992;116:349–57.
33. Sanchez-Fauquier A, Carrascosa AL, Carrascosa JL, Otero A, Glass RI, Lopez JA, et al. Characterization of a human astrovirus serotype 2 structural protein (VP26) that contains an epitope involved in virus neutralization. Virology. 1994;201:312–20.
34. Settembre EC, Chen JZ, Dormitzer PR, Grigorieff N, Harrison SC. Atomic model of an infectious rotavirus particle. EMBO J. 2011;30:408–16.
35. Speroni S, Rohayem J, Nenci S, Bonivento D, Robel I, Barthel J, et al. Structural and biochemical analysis of human pathogenic astrovirus serine protease at 2.0 A resolution. J Mol Biol. 2009;387:1137–52.
36. Xing L, Li T-C, Mayazaki N, Simon MN, Wall JS, Moore M, et al. Structure of hepatitis E virion-sized particle reveals an RNA-dependent viral assembly pathway. J Biol Chem. 2010;285:33175–83.
37. Yamashita T, Mori Y, Miyazaki N, Cheng RH, Yoshimura M, Unno H, et al. Biological and immunological characteristics of hepatitis E virus-like particles based on the crystal structure. Proc Natl Acad Sci USA. 2009;106:12986–91.
38. Yin J, Bergmann EM, Cherney MM, Lall MS, Jain RP, Vederas JC, et al. Dual modes of modification of hepatitis A virus 3C protease by a serine-derived beta-lactone: selective crystallization and formation of a functional catalytic triad in the active site. J Mol Biol. 2005;354:854–71.

Chapter 4
Astrovirus Pathogenesis

Victoria Meliopoulos and Stacey Schultz-Cherry

Abstract Astrovirus represents a public health problem and is responsible for a large proportion of diarrhea and enteric illness in children, the elderly, and the immunocompromised. In spite of its impact on human health, little is known about astrovirus pathogenesis. There are many reasons for this including the lack of a suitable small animal model for infection. This chapter will summarize the current state of our knowledge on astrovirus pathogenesis.

Introduction

While astroviruses are recognized as a leading cause of enteritis in infants, the elderly, and immunocompromised people, and are widespread in animals and birds, very little is known about astrovirus pathogenesis. Unlike rotavirus, there is minimal literature on the histopathological features and pathogenesis of astrovirus infection in humans or animals. This can be attributed to the lack of a suitable small animal model for astrovirus disease and difficulty in the propagation of astrovirus in vitro. In this chapter, we will describe the current literature concerning astrovirus pathogenesis in humans and animals and conclude with discussion and future studies focused on understanding the mechanism(s) of astrovirus-induced disease.

V. Meliopoulos, Ph.D. • S. Schultz-Cherry, Ph.D. (✉)
Department of Infectious Diseases, St. Jude Children's Research Hospital,
262 Danny Thomas Place, Memphis, TN 38105, USA
e-mail: Victoria.Meliopoulos@stjude.org; stacey.schultz-cherry@stjude.org

S. Schultz-Cherry (ed.), *Astrovirus Research: Essential Ideas, Everyday Impacts, Future Directions*, DOI 10.1007/978-1-4614-4735-1_4,

Astrovirus Disease in Humans

Human astroviruses (HAstVs) predominately affect children under the age of 2, elderly people, and immunocompromised individuals and are thought to be involved in up to 20% of the cases of sporadic nonbacterial diarrhea and 0.5–15% of diarrheal outbreaks [32, 50]. Astrovirus is most likely transmitted by the fecal–oral route [46]. Upon infection, clinical symptoms normally last between 2 and 4 days and consist mainly of watery diarrhea and less commonly vomiting, headache, fever, abdominal pain, and anorexia [18, 19, 40, 48]. However, there is evidence that HAstV infections can be asymptomatic in children and adults [45]. Astrovirus infection is characterized by blunting of the microvilli at the tips and changes in barrier function of the intestinal epithelium [27].

Until 2008, human infections were thought to be limited to eight closely related HAstV (HAstV-1–HastV-8) genotypes now referred to as the "classic" human astroviruses; although, there is a great deal of diversity within each of the different genotypes. In the early 2000s, two studies suggested that a variant could be considered a new subtype if identity was less than 95% nucleotide homology to a reference strain and >0.05 distance by phylogenetic analysis at the 3′-end of ORF2 [30, 62]. Others proposed that strains showing a sequence diversity of at least 7% could be considered as new lineages [24, 44]. To date there is no information on the association between disease symptoms and HAstV genotype, except for one study which suggests that HAstV-3 may be associated to more severe gastroenteritis and higher fecal viral titers when compared to the other genotypes [7].

The advent of pyrosequencing and pathogen discovery demonstrated that human stool contained a variety of previously unrecognized astroviruses including MLB1 (AstV-MLB1) and MLB-2 [15, 28], which are widely distributed and exhibit a great deal of heterogeneity within the group [15, 16] and AstV-VA1 [17]. Three novel and phylogenetically related human astroviruses were isolated from adult travelers, and were closely related to mink and ovine strains [31]. These novel strains were named human-, mink-, and ovine-like (HMOAstV) types A, B, and C, with HMOAstV-C being closely related to AstV-VA1 [31]. Based on these data, it is evident that there are at least three clades of astroviruses co-circulating in humans: classic, AstV-MLB1-like, and VA1-like (Fig. 4.1). Coinfection by two different genotypes in one person provides an opportunity for recombination to occur, and indeed, there has been evidence of recombination within the classic HAstV strains [62, 64]. It would be interesting to determine if asymptomatic infection correlates with a specific viral attribute, such as astrovirus genotype, or a host attribute like preexisting antibody.

Histological Change

In individuals with symptomatic HAstV infection, viral particles were detected in the low villous epithelium, in surface epithelium of a flat biopsy, and in macrophages of the lamina propria [47]. Infection appeared to be restricted to the villous enterocytes and exposed epithelium [53]. Sebire et al. recently described HAstV histopathology

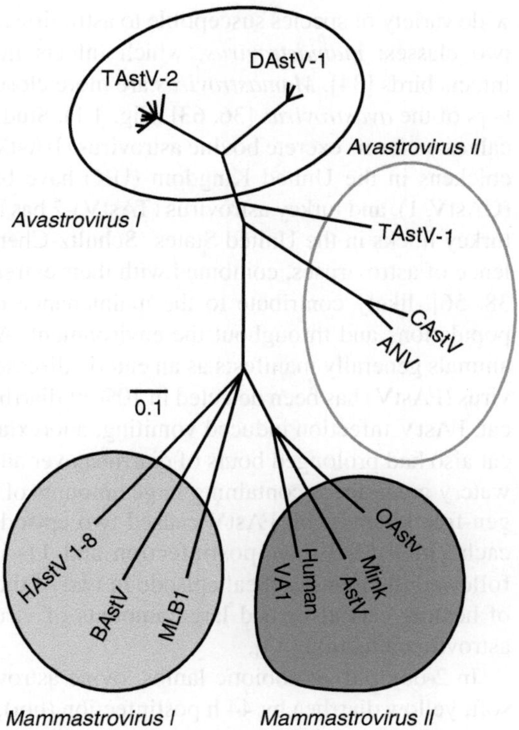

Fig. 4.1 Phylogenetic tree of representative whole genome of avian and mammalian astroviruses. The complete nucleotide sequence of the astrovirus genomes was aligned using MEGA4 software utilizing the Neighbor-Joining method and a Kimura 2-parameter model with pairwise deletion

in a human bone marrow transplant recipient afflicted with severe HAstV-induced diarrhea [57]. In this patient, replication was restricted to the apical two-thirds of enterocytes lining the villi, with a progressive increase in viral staining throughout the small intestine. There was evidence of small intestinal, but not gastric involvement by astrovirus, with more extensive infection in the jejunum as compared to the duodenum. Although very limited data exist, this study suggests that despite the severe diarrhea, astrovirus infection is associated with relatively minor and nonspecific morphological changes in the intestines of infected individuals. In addition, the inflammatory response was mild in spite of good evidence of engraftment, and other acute and chronic inflammatory reactions have been noted in immunocompromised patients [57]. These results affirm the apparently small contribution of histological damage and inflammation to astrovirus-induced diarrhea.

Astrovirus Disease in Animals and Birds

Astroviruses are common in nature and have so far been identified in a variety of species, including minks [14], chickens [1, 29], cats [25], dogs [42], mice [33], deer [61], ducks [20], turkeys [37], lambs [21, 25, 59], and calves [65]. Because of the wide variety of species susceptible to astrovirus, astroviruses have been divided into two classes: *mamastrovirus*, which infects mammals, and *avastrovirus*, which infects birds [34]. *Mamastrovirus* are more closely related to each other than members of the *avastrovirus* [36, 63] (Fig. 4.1). Studies have indicated that 60–100% of calves on farms excrete bovine astrovirus (BAstV) [6]. A large proportion of farmed chickens in the United Kingdom (UK) have been exposed to chicken astrovirus (CAstV, 1), and turkey astrovirus (TAstV)-2 has been isolated from 78% of diarrheic turkey flocks in the United States (Schultz-Cherry, unpublished result). The prevalence of astroviruses, combined with their extreme environmental stability [26, 34, 38, 56], likely contribute to the maintenance of virus within animal and human populations and throughout the environment. As in humans, astrovirus disease in animals generally manifests as an enteric disease although not always. Feline astrovirus (FAstV) has been detected in 10% of diarrheic cats [43]. In a naturally infected cat, FAstV infection induced vomiting, anorexia, dehydration, and depression. The cat also had prolonged bouts of diarrhea over an 8-week period, which consisted of watery green feces containing large amounts of FAstV. Infection of specific pathogen-free kittens with FAstV caused two episodes of pyrexia and mild diarrhea in each kitten (2–3 days postinfection and 11–12 days postinfection). Viral shed followed the first diarrheal episode in two of the three kittens [25]. In contrast, 3% of healthy cats also shed large amounts of virus while showing no symptoms of astrovirus infection [43].

In 2-day-old gnotobiotic lambs, ovine astrovirus (OAstV) infection resulted in soft, yellow diarrhea by 44 h postinfection (hpi), in contrast to the firm, dark brown feces of control lambs. Shedding of OAstV was detected from 38 to 120 h postinfection [58]. Porcine astroviruses (PAstV) have been identified in association with rotavirus and calicivirus in young piglets. Subsequently, infected piglets developed similar symptoms, including anorexia, diarrhea, and a failure to thrive. Infection led to the death of approximately 15% of infected piglets [5]. Astrovirus-induced diarrhea has also been identified in deer [61] and mink [14], although disease has not been well described in these animals.

Turkey astrovirus (TAstV) was originally identified in the feces of young (6–11 days of age) turkeys with a history of diarrhea [55]. Since then, two distinct genotypes, TAstV-1 and TAstV-2, have been identified [8, 36]. Turkeys infected with TAstV-2 developed a profuse, watery diarrhea by 24 h postinfection, which continued until 12 days postinfection [35]. Morphologically, the intestines of infected poults were three to five times larger than their control counterparts and appeared dilated, distended, and fluid-filled by 3 days postinfection [2]. TAstV-2 infection resulted in an overall growth depression of infected birds by 5 days postinfection and throughout the experiment, possibly as a consequence of decreased nutrient absorption [35]. Additionally, TAstV-2-infected turkeys demonstrated a transient, localized growth depression of the thymus, which manifested by 5 days postinfection and resolved by 12 days postinfection [35]. Finally, TAstV-2 was shown to cause viremia, and infectious virus was isolated throughout the tissues including the thymus, bursa, spleen, liver, kidney, pancreas, skeletal muscle, plasma, and bone marrow [35], although the virus did not appear to replicate outside the small intestine [2]. The importance of this finding in pathogenesis is under investigation. However, given the recent reports of systemic spread and viremia in humans and neurological disease in infected mink, it is likely that certain astroviruses are capable of inducing viremia and causing systemic spread. This would suggest that the systemic spread with TAstV-2 is not unique to the turkey model. Similarly, astrovirus in chickens resulted in mild diarrhea by 4 days postinfection; however, growth depression correlating with infection was not observed [1].

As mentioned previously, astrovirus infection in animals does not manifest exclusively as gastroenteritis. In 1984, Gough et al. isolated astrovirus from ducks (DAstV) dying from severe hepatitis. DAstV was identified in the liver as well as in the feces. In field studies, DAstV caused an age-dependent mortality; 10–25% of ducks 4–6 weeks of age died of astrovirus-induced hepatitis, while ducks 6–14 days of age exhibited mortality rates up to 50%. Experimental DAstV infection of ducks 2–3 days of age with isolated DAstV resulted in 25% mortality by 4 days postinfection. Serial passage of DAstV through ducks yielded mortality rates up to 55% [20]. Additionally, avian nephritis virus (ANV), recently identified as an astrovirus, is an etiological agent of interstitial nephritis, causing growth retardation of young chickens [29].

As in the case of human astroviruses, some animal astroviruses do not cause any detectable disease. BAstV infection, for example, is asymptomatic [6] or induces only slight color and consistency changes in the feces of gnotobiotic calves 1–42 days of age [65–67]. Despite viral shed from 5 to 8 days postinfection, diarrhea was never detected in infected calves [6]. Canine astrovirus infection can also be asymptomatic and has been detected infrequently in naturally infected dogs, none of which had diarrhea [42]. Even astroviruses that have been documented to cause disease occasionally do not. The reason for this is unclear although it could be related to age of the infected host, strain of virus, or even level of viral replication.

The discrepancy in astrovirus severity is poorly characterized. The absence of discernable disease in gnotobiotic calves suggests that BAstV by itself is not pathogenic. However, further work by Woode et al. demonstrated that astrovirus infection with other enteric pathogens, such as Breda virus and rotavirus, increased the severity of astrovirus infection. A larger proportion of astrovirus-infected cells were detected, and intestinal lesions were more consistently present in coinfected animals. Additionally, coinfection always led to diarrhea, although titers of shed BAstV did not change. Titers of the coinfected virus were not monitored [67]. This finding has significant implications, considering that astrovirus is often isolated with other enteric pathogens, both in animals and in humans [1, 3, 10–12, 20, 22, 41, 43]. If this observation holds true for all astroviruses, coinfections may result in a synergistically more serious disease in naturally infected animals. Further work is necessary to clarify the effect of coinfection on astrovirus pathogenesis.

Histological Changes

While astrovirus infection has been demonstrated in a variety of species, thorough pathogenesis studies have been limited to lambs, calves, and turkeys. In each of these model systems, astrovirus replication has been demonstrated in the small intestine, although the exact pattern of infection and clinical manifestations of infection differs between animals.

Lambs

Early investigation into astrovirus pathogenesis was conducted in gnotobiotic lambs [58]. Infection was apparent in the jejunum and ileum of the small intestine by 14 h postinfection at which time infection was localized to the enterocytes lining the tips of the villi. By 23 h postinfection, most of the enterocytes lining the upper two-thirds of the villi were infected. Infection of crypt cells was never observed [21]. Morphologically, infected enterocytes appeared highly vacuolated, with large basal vacuoles as well as multiple smaller apical vacuoles. While uninfected villi had smooth edges, the margins of infected villi appeared ragged due to the presence of microcrypts [58]. Viral replication peaked at 38 h postinfection [58], at which time infection had spread throughout the intestine. This correlated with production of virus, which was identified in the cytoplasm as well as various membrane-bound vesicles. OAstVs also appear to line the outer edges of the villi, although release of the virus was never observed [21]. By 38 h postinfection, a mild villous atrophy was observed, which correlated with sloughing of degenerate enterocytes into the intestinal lumen. Replacement of these cells with cuboidal enterocytes corresponded with clearance of OAstV, which was never identified past 38 h postinfection [21]. Regrowth of cuboidal cells, villous atrophy, and an increase in goblet cells persisted through 70 h postinfection [21]. By 120 h postinfection, intestines from OAstV-infected lambs were indistinguishable from control lambs [21, 58].

Calves

A somewhat different pattern of infection was observed in gnotobiotic calves infected with BAstV. In these animals, replication was identified at 1–3 days postinfection and was localized to dome enterocytes overlying Peyers patches. Occasionally, infection spread to the base of nearby villi. Mild lesions, consisting of flat to cuboidal enterocytes localized to intestinal domes were also found [67]. A slight villous atrophy in addition to thickening of the villi was noted in some animals, and viral shedding was demonstrated up to 5 days postinfection [65].

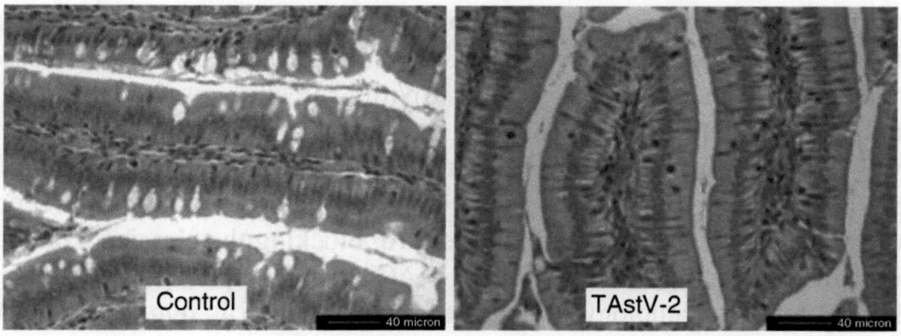

Fig. 4.2 TastV-2 infection is not associated with cytopathic changes in the intestines of infected turkey poults. Duodenal sections were collected from control (**a**) or TAstV-2 (**b**) turkey poults 3 days postinfection and stained by hematoxylin and eosin. Magnification is given in the *bar*

Turkeys

TAstV-2 replication was restricted to the intestines as determined by in situ hybridization. By 24 h postinfection, replication was detected in the upper regions of the small intestine [2]. Individual degenerating enterocytes were apparent along the basal edge of villi by 2 days postinfection and continued through day 4. By day 5, mild shortening of the villi was observed, as well as occasional clusters of necrotic enterocytes along the villous base, correlating with infection [35]. By 7 days postinfection, TAstV-2 was localized primarily to the large intestine. Replication peaked at 3–5 days postinfection, and TAstV-2 was infrequently observed at time points later than 9 days postinfection [2]. Although replication was limited to the intestine, infectious TAstV-2 was isolated from the bursa, thymus, spleen, kidney, skeletal muscle, pancreas, and plasma [35]. Overall, TAstV-2 infection was associated with mild histopathology (Fig. 4.2) and little cell death as measured by TUNEL [35]. The lack of an inflammatory response may result from the increase in transforming growth factor-β (TGF-β) in infected intestines [35]. Intriguingly, the histopathology observed in the young turkey model is similar to that observed in human infection [57]. These results suggest that astrovirus-induced diarrhea is due to a mechanism other than destruction of the intestinal epithelium or eliciting an inflammatory response, and raise the question of how astroviruses induce diarrhea. Although we do not have an answer, there are several possible mechanisms that warrant discussion.

Mechanism of Astrovirus-Induced Diarrhea

In spite of the recent progress, the mechanism of astrovirus-induced diarrhea remains unknown. Astroviruses could induce diarrhea through several different mechanisms, including destruction of the intestinal epithelia, modulation of ion channels or intestinal disaccharidase activity leading to malabsorption, or induction of intestinal epithelial barrier permeability. A recent study demonstrated that HAstV-4 induced apoptosis in infected Caco-2 cells [23]. However, increased cell death has not been associated with astrovirus infection in vivo [35, 50]. An inflammatory response is most likely not responsible for cellular damage, as levels of TGF-β, an anti-inflammatory cytokine, were increased in astrovirus-infected turkeys [35]. Further immunosuppressive activity has been noted in a report that the astrovirus capsid protein can bind the C1q, blocking the classical complement pathway and preventing formation of the MAC attack complex [4].

The lack of histopathology during astrovirus infection is reminiscent of *Vibrio cholera*-induced diarrhea. Toxin-induced diarrhea can occur through direct effects on ion transport in intestinal epithelial cells (i.e., direct toxin interaction with intrinsic ion channels in the membrane) or through indirect interactions with ion transport in intestinal epithelial cells mediated by toxin binding to a membrane receptor. These effects consequently cause the release of and/or changes in second messengers that affect calcium and potassium permeable channels [39]. Further studies examining modulation of ion concentration during astrovirus infection could yield interesting results.

Evidence for the Role of Malabsorption

Studies of TAstV-1 in a turkey model demonstrated that specific maltase activity (SMA) indicating intestinal absorption decreased during infection. SMA was significantly lower in astrovirus-inoculated poults than in control birds throughout the entire small intestine from day 3 to day 7 postinfection. Decreased SMA resulted in disaccharide maldigestion, malabsorption, and subsequent osmotic diarrhea [60]. More recently, studies in TAstV-2-infected turkey poults revealed reduced mucosal-to-serosal sodium ion flux and decreased sodium absorption in the intestine [52]. Further investigation determined that the sodium hydrogen exchangers NHE2 and NHE3, proteins located on the brush border of the intestinal epithelium and involved in sodium absorption [13, 68], were altered upon TAstV-2 infection [52]. NHE2 levels were increased upon infection and NHE3 apical localization was disrupted [52]. These findings suggest that astrovirus-induced sodium malabsorption may be involved in the induction of diarrhea, at least in the turkey model of infection.

Evidence for Role of Dysregulated Intestinal Permeability

An alternative, although not mutually exclusive, hypothesis is that astroviruses modulate the permeability of intestinal epithelial barrier. Epithelial cells, including the intestinal enterocytes, form a tight barrier to the migration of fluids and solutes into and out of the intestine through the interactions of specific proteins known collectively as the apical junctional complex (Fig. 4.3a). These proteins are solely responsible for maintaining the barrier function of the intestine and disruption of

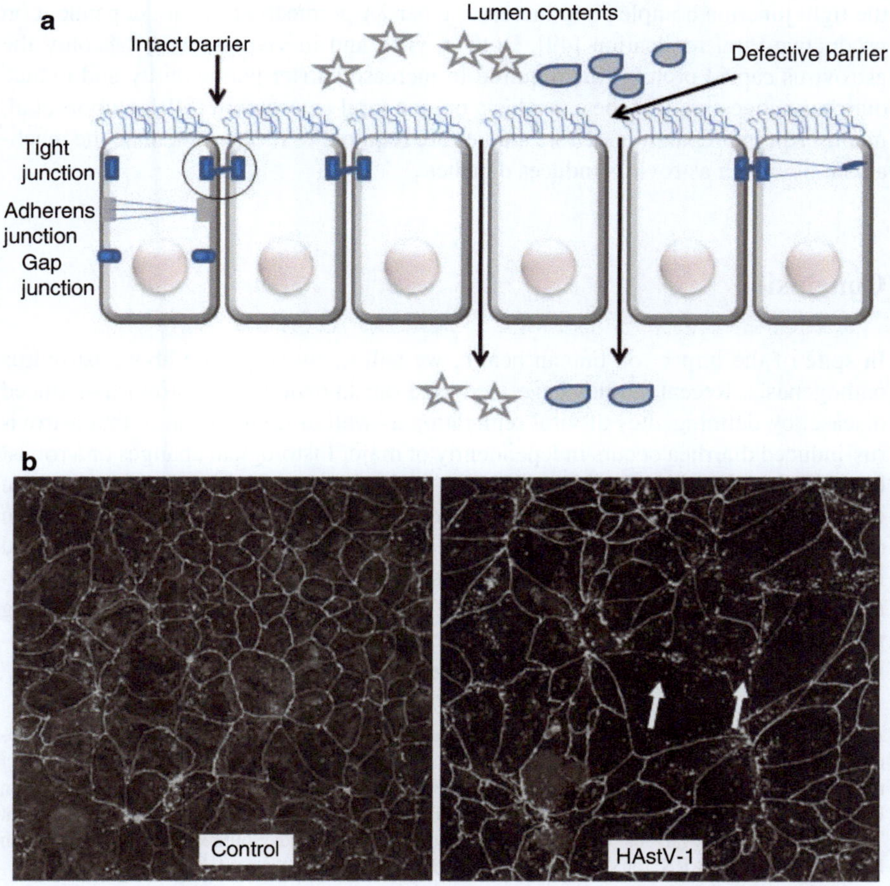

Fig. 4.3 HAstV-1 disrupts intestinal epithelial cell tight junction by reorganizing occludin. (**a**) Schematic of intestinal epithelial cells. The junctional complex containing occludin, which is located at the most apical region of lateral membranes, is *circled*. (**b**) Confocal microscopy image of uninfected (control) and HAstV-1-infected (HAstV-1) differentiated Caco-2 cells stained for occludin 24 h postinfection. *Arrows* indicate areas where the cobblestone staining is disrupted due to relocalization of occludin from the periphery of HAstV-1-infected cells

this barrier can result in diarrhea [9]. Not much is known regarding how astroviruses modulate barrier permeability, although preliminary studies are beginning to give insight into the process. During TAstV-2 infection of turkey poults, virus-induced F-actin rearrangement was noted possibly leading to disruption of ionic permeability and reduced sodium ion influx [52]. In vitro, HAstV was shown to increase the permeability of differentiated Caco-2 intestinal epithelial cells in a time and dose-dependent but HAstV subtype-independent manner [49, 50]. This appeared to be caused by F-actin rearrangement and occludin redistribution (Fig. 4.3b), suggesting a possible mechanism for systemic spread [49]. Intriguingly, the reorganization of the tight junction complex and increase in barrier permeability is not dependent on productive viral replication [49]. Both in vitro and in vivo showed that only the astrovirus capsid protein was required to increase barrier permeability and induce diarrhea suggesting that the capsid is a unique viral enterotoxin ([49]; Dubois et al. manuscript in preparation). More studies are required to further elucidate the mechanism by which astrovirus induces diarrhea.

Conclusion

In spite of the impact on human health, we still know very little about astrovirus pathogenesis. Recent studies have increased our knowledge of astrovirus-induced disease, by defining sites of viral replication as well as demonstrating that astrovirus-induced diarrhea occurs independently of major histological changes or a robust host inflammatory response. Further work is required to understand the mechanism of tight junction dysregulation caused by astrovirus and to understand how infection can lead to systemic spread. Astrovirus is of particular concern to the pediatric and immunocompromised population, thus it is crucial to define and understand astrovirus pathogenesis. Ongoing studies in our laboratory are focused on understanding the mechanism of astrovirus-induced diarrhea using both in vitro (intestinal epithelial cells) and in vivo model systems to yield insight into astrovirus disease.

Acknowledgments We would like to thank past and present members of the Schultz-Cherry laboratory for their ongoing discussions and support; and the following financial sources throughout the years: U.S. Poultry & Egg Association, Hartwell Foundation, the St. Jude Children's Infection Defense Center funds, and ALSAC. Also, we acknowledge the generous support of the National Turkey Federation, various commercial turkey producers, and Southeast Poultry Research Laboratory, USDA-ARS, for supplying commercial turkey poults.

References

1. Baxendale W, Mebatsion T. The isolation and characterisation of astroviruses from chickens. Avian Pathol. 2004;33:364–70.
2. Behling-Kelly E, Schultz-Cherry S, Koci M, Kelley L, Larsen D, Brown C. Localization of astrovirus in experimentally infected turkeys as determined by in situ hybridization. Vet Pathol. 2002;39:595–8.
3. Bon F, Fascia P, Dauvergne M, Tenenbaum D, Planson H, Petion AM, et al. Prevalence of group A rotavirus, human calicivirus, astrovirus, and adenovirus type 40 and 41 infections among children with acute gastroenteritis in Dijon, France. J Clin Microbiol. 1999;37:3055–8.
4. Bonaparte RS, Hair PS, Banthia D, Marshall DM, Cunnion KM, Krishna NK. Human astrovirus coat protein inhibits serum complement activation via C1, the first component of the classical pathway. J Virol. 2008;82:817–27.
5. Bridger JC. Detection by electron microscopy of caliciviruses, astroviruses and rotavirus-like particles in the faeces of piglets with diarrhoea. Vet Rec. 1980;107:532–3.

6. Bridger JC, Hall GA, Brown JF. Characterization of a calici-like virus (Newbury agent) found in association with astrovirus in bovine diarrhea. Infect Immun. 1984;43:133–8.
7. Caballero S, Guix S, El-Senousy WM, Calico I, Pinto RM, Bosch A. Persistent gastroenteritis in children infected with astrovirus: association with serotype-3 strains. J Med Virol. 2003;71:245–50.
8. Calnek BW, Barnes HJ, Beard CW, editors. Diseases of poultry. 11th ed. Ames: Iowa State University Press; 2003.
9. Clayburgh DR, Shen L, Turner JR. A porous defense: the leaky epithelial barrier in intestinal disease. Lab Invest. 2004;84:282–91.
10. Cunliffe NA, Dove W, Gondwe JS, Thindwa BD, Greensill J, Holmes JL, et al. Detection and characterisation of human astroviruses in children with acute gastroenteritis in Blantyre, Malawi. J Med Virol. 2002;67:563–6.
11. Dalton RM, Roman ER, Negredo AA, Wilhelmi ID, Glass RI, Sanchez-Fauquier A. Astrovirus acute gastroenteritis among children in Madrid, Spain. Pediatr Infect Dis J. 2002;21:1038–41.
12. Dennehy PH, Nelson SM, Spangenberger S, Noel JS, Monroe SS, Glass RI. A prospective case-control study of the role of astrovirus in acute diarrhea among hospitalized young children. J Infect Dis. 2001;184:10–5.
13. Donowitz M, De La Horra C, Calonge ML, Wood IS, Dyer J, Gribble SM, et al. In birds, NHE2 is major brush-border Na+/H+exchanger in colon and is increased by a low-NaCl diet. Am J Physiol. 1998;274:R1659–69.
14. Englund L, Chriel M, Dietz HH, Hedlund KO. Astrovirus epidemiologically linked to pre-weaning diarrhoea in mink. Vet Microbiol. 2002;85:1–11.
15. Finkbeiner SR, Kirkwood CD, Wang D. Complete genome sequence of a highly divergent astrovirus isolated from a child with acute diarrhea. Virol J. 2008;5:117.
16. Finkbeiner SR, Le BM, Holtz LR, Storch GA, Wang D. Detection of newly described astrovirus MLB1 in stool samples from children. Emerg Infect Dis. 2009;15:441–4.
17. Finkbeiner SR, Li Y, Ruone S, Conrardy C, Gregoricus N, Toney D, et al. Identification of a novel astrovirus (astrovirus VA1) associated with an outbreak of acute gastroenteritis. J Virol. 2009;83:10836–9.
18. Glass RI, Bresee J, Jiang B, Gentsch J, Ando T, Fankhauser R, et al. Gastroenteritis viruses: an overview. Novartis Found Symp. 2001;238:5–19. discussion 19–25.
19. Glass RI, Noel J, Mitchell D, Herrmann JE, Blacklow NR, Pickering LK, et al. The changing epidemiology of astrovirus-associated gastroenteritis: a review. Arch Virol Suppl. 1996;12:287–300.
20. Gough RE, Collins MS, Borland E, Keymer LF. Astrovirus-like particles associated with hepatitis in ducklings. Vet Rec. 1984;114:279.
21. Gray EW, Angus KW, Snodgrass DR. Ultrastructure of the small intestine in astrovirus-infected lambs. J Gen Virol. 1980;49:71–82.
22. Gray JJ, Wreghitt TG, Cubitt WD, Elliot PR. An outbreak of gastroenteritis in a home for the elderly associated with astrovirus type 1 and human calicivirus. J Med Virol. 1987;23:377–81.
23. Guix S, Bosch A, Ribes E, Dora L, Martinez, Pinto RM. Apoptosis in astrovirus-infected CaCo-2 cells. Virology. 2004;319:249–61.
24. Guix S, Caballero S, Villena C, Bartolome R, Latorre C, Rabella N, et al. Molecular epidemiology of astrovirus infection in Barcelona, Spain. J Clin Microbiol. 2002;40:133–9.
25. Harbour DA, Ashley CR, Williams PD, Gruffydd-Jones TJ. Natural and experimental astrovirus infection of cats. Vet Rec. 1987;120:555–7.
26. Herring AJ, Gray EW, Snodgrass DR. Purification and characterization of ovine astrovirus. J Gen Virol. 1981;53:47–55.
27. Hodges K, Gill R. Infectious diarrhea: cellular and molecular mechanisms. Gut Microbes. 2010;1:4–21.
28. Holtz LR, Wylie KM, Sodergren E, Jiang Y, Franz CJ, Weinstock GM, et al. Astrovirus MLB2 viremia in febrile child. Emerg Infect Dis. 2011;17:2050–2.

29. Imada T, Yamaguchi S, Mase M, Tsukamoto K, Kubo M, Morooka A. Avian nephritis virus (ANV) as a new member of the family Astroviridae and construction of infectious ANV cDNA. J Virol. 2000;74:8487–93.
30. Jakab F, Walter JE, Berke T, Matson DO, Mitchell DK, Szucs G. Molecular characterization and sequence analysis of human astroviruses circulating in Hungary. FEMS Immunol Med Microbiol. 2003;39:97–102.
31. Kapoor A, Li L, Victoria J, Oderinde B, Mason C, Pandey P, et al. Multiple novel astrovirus species in human stool. J Gen Virol. 2009;90:2965–72.
32. Kirkwood CD, Clark R, Bogdanovic-Sakran N, Bishop RF. A 5-year study of the prevalence and genetic diversity of human caliciviruses associated with sporadic cases of acute gastroenteritis in young children admitted to hospital in Melbourne, Australia (1998-2002). J Med Virol. 2005;77:96–101.
33. Kjeldsberg E, Hem A. Detection of astroviruses in gut contents of nude and normal mice. Brief report. Arch Virol. 1985;84:135–40.
34. Knipe DM, Howley PM, editors. Fields virology, vol. 1. Philadelphia: Lippincott Williams & Wilkins; 2001.
35. Koci MD, Moser LA, Kelley LA, Larsen D, Brown CC, Schultz-Cherry S. Astrovirus induces diarrhea in the absence of inflammation and cell death. J Virol. 2003;77:11798–808.
36. Koci MD, Schultz-Cherry S. Avian astroviruses. Avian Pathol. 2002;31:213–27.
37. Koci MD, Seal BS, Schultz-Cherry S. Molecular characterization of an avian astrovirus. J Virol. 2000;74:6173–7.
38. Kurtz JB, Lee TW, Parsons AJ. The action of alcohols on rotavirus, astrovirus and enterovirus. J Hosp Infect. 1980;1:321–5.
39. Laohachai KN, Bahadi R, Hardo MB, Hardo PG, Kourie JI. The role of bacterial and non-bacterial toxins in the induction of changes in membrane transport: implications for diarrhea. Toxicon. 2003;42:687–707.
40. LeBaron CW, Furutan NP, Lew JF, Allen JR, Gouvea V, Moe C, et al. Viral agents of gastroenteritis. Public health importance and outbreak management. MMWR Recomm Rep. 1990;39: 1–24.
41. Maldonado Y, Cantwell M, Old M, Hill D, Sanchez ML, Logan L, et al. Population-based prevalence of symptomatic and asymptomatic astrovirus infection in rural Mayan infants. J Infect Dis. 1998;178:334–9.
42. Marshall JA, Healey DS, Studdert MJ, Scott PC, Kennett ML, Ward BK, et al. Viruses and virus-like particles in the faeces of dogs with and without diarrhoea. Aust Vet J. 1984;61: 33–8.
43. Marshall JA, Kennett ML, Rodger SM, Studdert MJ, Thompson WL, Gust ID. Virus and virus-like particles in the faeces of cats with and without diarrhoea. Aust Vet J. 1987;64:100–5.
44. Medina SM, Gutierrez MF, Liprandi F, Ludert JE. Identification and type distribution of astroviruses among children with gastroenteritis in Colombia and Venezuela. J Clin Microbiol. 2000;38:3481–3.
45. Mendez-Toss M, Griffin DD, Calva J, Contreras JF, Puerto FI, Mota F, et al. Prevalence and genetic diversity of human astroviruses in Mexican children with symptomatic and asymptomatic infections. J Clin Microbiol. 2004;42:151–7.
46. Midthun K, Greenberg HB, Kurtz JB, Gary GW, Lin FY, Kapikian AZ. Characterization and seroepidemiology of a type 5 astrovirus associated with an outbreak of gastroenteritis in Marin County, California. J Clin Microbiol. 1993;31:955–62.
47. Mitchell DK. Astrovirus gastroenteritis. Pediatr Infect Dis J. 2002;21:1067–9.
48. Mitchell DK, Matson DO, Jiang X, Berke T, Monroe SS, Carter MJ, et al. Molecular epidemiology of childhood astrovirus infection in child care centers. J Infect Dis. 1999;180: 514–7.
49. Moser LA, Carter M, Schultz-Cherry S. Astrovirus increases epithelial barrier permeability independently of viral replication. J Virol. 2007;81:11937–45.
50. Moser LA, Schultz-Cherry S. Pathogenesis of astrovirus infection. Viral Immunol. 2005; 18:4–10.

51. Nighot PK, Moeser A, Ali RA, Blikslager AT, Koci MD. Astrovirus infection induces sodium malabsorption and redistributes sodium hydrogen exchanger expression. Virology. 2010; 401:146–54.
52. Phillips AD, Rice SJ, Walker-Smith JA. Astrovirus within human small intestinal mucosa. Gut. 1982;23:A923–4.
53. Reynolds DL, Saif YM. Astrovirus: a cause of an enteric disease in turkey poults. Avian Dis. 1986;30:728–35.
54. Schultz-Cherry S, King DJ, Koci MD. Inactivation of an astrovirus associated with poult enteritis mortality syndrome. Avian Dis. 2001;45:76–82.
55. Sebire NJ, Malone M, Shah N, Anderson G, Gaspar HB, Cubitt WD. Pathology of astrovirus associated diarrhoea in a paediatric bone marrow transplant recipient. J Clin Pathol. 2004; 57:1001–3.
56. Snodgrass DR, Angus KW, Gray EW, Menzies JD, Paul G. Pathogenesis of diarrhoea caused by astrovirus infections in lambs. Arch Virol. 1979;60:217–26.
57. Snodgrass DR, Gray EW. Detection and transmission of 30 nm virus particles (astroviruses) in faeces of lambs with diarrhoea. Arch Virol. 1977;55:287–91.
58. Thouvenelle ML, Haynes JS, Sell JL, Reynolds DL. Astrovirus infection in hatchling turkeys: alterations in intestinal maltase activity. Avian Dis. 1995;39:343–8.
59. Tzipori S, Menzies JD, Gray EW. Detection of astrovirus in the faeces of red deer. Vet Rec. 1981;108:286.
60. Walter JE, Briggs J, Guerrero ML, Matson DO, Pickering LK, Ruiz-Palacios G, et al. Molecular characterization of a novel recombinant strain of human astrovirus associated with gastroenteritis in children. Arch Virol. 2001;146:2357–67.
61. Willcocks MM, Ashton N, Kurtz JB, Cubitt WD, Carter MJ. Cell culture adaptation of astrovirus involves a deletion. J Virol. 1994;68:6057–8.
62. Wolfaardt M, Kiulia NM, Mwenda JM, Taylor MB. Evidence of a recombinant wild-type human astrovirus strain from a Kenyan child with gastroenteritis. J Clin Microbiol. 2011;49: 728–31.
63. Woode GN, Bridger JC. Isolation of small viruses resembling astroviruses and caliciviruses from acute enteritis of calves. J Med Microbiol. 1978;11:441–52.
64. Woode GN, Gourley NE, Pohlenz JF, Liebler EM, Mathews SL, Hutchinson MP. Serotypes of bovine astrovirus. J Clin Microbiol. 1985;22:668–70.
65. Woode GN, Pohlenz JF, Gourley NE, Fagerland JA. Astrovirus and Breda virus infections of dome cell epithelium of bovine ileum. J Clin Microbiol. 1984;19:623–30.
66. Zachos NC, Tse M, Donowitz M. Molecular physiology of intestinal Na+/H+exchange. Annu Rev Physiol. 2005;67:411–43.

Chapter 5
Immune Responses

Neel K. Krishna, Matthew D. Koci, and Susana Guix

Abstract Members of the *Astroviridae* family are recognized to be a leading cause of viral enteritis in infants, the immune-compromised, and elderly, as well as a significant cause of disease in other mammals and avian species. However, little is known of the immune response to astrovirus infection. The humoral immune response is believed to restrict infection and clinical disease in humans as documented by epidemiological and clinical observational studies. Consistent with the concept that neutralizing antibodies may control astrovirus infection, limited studies have detected the presence of astrovirus-specific $CD4^{+}$ helper T cells in human volunteers. However, no other data regarding cell-mediated immunity has been reported, and some small animal models have recently highlighted the important role of innate immune responses in controlling astrovirus infections, especially primary infections in young animals. Consequently, research in the past decade has focused on the role of the innate immune system in the anti-astrovirus response. In this chapter, we summarize the current knowledge of the immune response to astrovirus from studies on normal healthy adults as well as the young, elderly, and immune-compromised. In addition, we focus on findings from a small animal model that implicate a role for the innate immune response in astrovirus immunity. Finally,

N.K. Krishna, Ph.D. (✉)
Department of Microbiology and Molecular Cell Biology, Eastern Virginia Medical School, 700 West Olney Road, Norfolk, VA 23507, USA
e-mail: krishnnk@evms.edu

M.D. Koci, Ph.D.
Department of Poultry Science, North Carolina State University, Campus Box 7608, Raleigh, NC 27695, USA
e-mail: mdkoci@ncsu.edu

S. Guix, Ph.D.
Enteric Virus Laboratory, Department of Microbiology, Institute of Nutrition and Food Safety, University of Barcelona, Avda. Diagonal 643, Barcelona 08028, Spain
e-mail: susanaguix@ub.edu

S. Schultz-Cherry (ed.), *Astrovirus Research: Essential Ideas, Everyday Impacts, Future Directions*, DOI 10.1007/978-1-4614-4735-1_5,

we review recent studies that demonstrate an unexpected interaction of complement system proteins with the human astroviruses that may have implications for astrovirus immune evasion and replication in humans.

Introduction

Although astroviruses are recognized to be a leading cause of viral enteritis in infants, the elderly, and the immune-compromised, as well as a significant cause of disease in other mammals and avian species, very little is known regarding the immune response to astrovirus infection. Epidemiological and clinical observational studies have demonstrated that the humoral immune response plays a role in restraining infection and clinical disease in humans. Consistent with the concept that neutralizing antibodies may control astrovirus infection, limited studies have detected the presence of astrovirus-specific $CD4^+$ helper T cells in human volunteers. However, no other data regarding cell-mediated immunity have been reported and some small animal models have recently highlighted the important role of innate immune responses in controlling astrovirus infections, especially primary infections in young animals. Consequently, research in the past decade has focused on the role of the innate immune system in the anti-astrovirus response. In this chapter, we summarize the current knowledge of the immune response to astrovirus from studies on normal healthy adults as well as the young, elderly, and immune-compromised. In addition, we focus on findings from a small animal model that implicate a role for the innate immune response in astrovirus immunity. Finally we review recent studies that demonstrate an unexpected interaction of complement system proteins with the human astroviruses that may have implications for astrovirus immune evasion and replication in humans.

Normal Healthy Adults

Human astroviruses (HAstVs) are largely regarded as a disease of the very young, the elderly, and the immune-compromised with very few reports of astrovirus-mediated disease in normal healthy adults. The endemic nature of astrovirus disease and the biphasic age distribution has led to the hypothesis that the adaptive immune system protects adults from infection. This idea is based largely on epidemiological studies demonstrating that more than 70% of healthy young adults have anti-astrovirus antibodies [36]. These findings correlate with observations made during two different astrovirus clinical studies using human volunteers. In these studies, the majority of infected subjects demonstrated mild clinical signs; however, a few subjects did demonstrate more severe disease [37]. These individuals were found to have no anti-astrovirus antibodies prior to infection, unlike the majority of the subjects that demonstrated less severe signs [37, 47].

In addition to the presence of anti-astrovirus antibodies, studies by Molberg et al. have demonstrated the presence of astrovirus-specific $CD4^+$ and $CD8^+$ T cells in intestinal biopsy samples from patients with no signs of infectious intestinal disorders [50]. While there are more reports correlating anti-astrovirus antibody levels and clinical signs, the identification of astrovirus-specific T cells in the intestine suggests that both aspects of the adaptive immune response are involved in protecting the normal healthy adult from astrovirus infections, and that physiological factors that impair these systems will result in increased risk of an astrovirus-mediated gastroenteritis.

The Young, Elderly, and Immune-Compromised

It is well documented that astrovirus gastroenteritis in humans and symptomatic disease in other mammalian and avian species has a specific tropism for the young [45]. In addition, clinical disease from astrovirus infection has also been documented in the elderly and immune-compromised. The biphasic age distribution of astrovirus-induced clinical disease suggests that neutralizing antibody attained early in life provides protection from symptomatic disease during adult life and that astrovirus immunity wanes in later life [45]. To date, there have been no reports in the literature to suggest why astroviruses (or other viral pathogens such as rotavirus and caliciviruses) demonstrate a predilection to infecting the young other than the general immaturity of the developing immune system. Various studies have demonstrated that antibodies to astrovirus are prevalent in serum pools around the world, which would be expected given the endemic nature of human astroviruses [7, 12, 34, 36, 54]. The observation that preterm and term neonates are susceptible to astrovirus infection [1, 2, 63] along with infants younger than 1 year of age [47, 62] suggests that maternally acquired antibody may not necessarily be protective for astrovirus infection.

Outbreaks of human astrovirus gastroenteritis have been reported in elderly, institutionalized individuals [25, 39, 43, 44, 55, 71]. Symptomatic infection of the elderly may be correlated with insufficient amounts of protective antibody due to waning antibody levels with increasing age. However, more studies are required to evaluate this hypothesis.

A number of studies have implicated human astroviruses as the causative agent of diarrhea in a variety of immune-compromised individuals (Table 5.1). Two recently described studies are worthy of note, as they implicate astrovirus as the causative agent of neurological disease in two unrelated immune-compromised conditions. The first involves the postmortem identification of astrovirus as the causative agent of encephalitis in a 15-year-old boy with X-linked agammaglobulinemia [57]. In this particular case, unbiased pyrosequencing identified a novel human astrovirus in CNS tissue. The full-length genome of the virus, tentatively named human astrovirus-Puget Sound (HAstV-PS) was determined. The capsid protein gene shows significant homology with human astrovirus VA1 and is closely related

Table 5.1 Astrovirus infection in immune-compromised patients

Immune deficiency	Reference
Chronic lymphocytic leukemia	[10]
Waldenstrom's macrogulinemia	[3]
Bone marrow transplantation	[11, 12, 61]
HIV infection	[13, 23, 27, 40]
Primary immunodeficiency	[20]
X-linked agammaglobulinemia	[57]
Transplantation/short bowel syndrome	[62]
T-cell immunodeficiency	[72]
Severe combined immunodeficiency	[75]

to mink astrovirus. Interestingly, the patient lived near a mink farm suggesting that exposure to mink might have been the potential source of infection. Although mink astroviruses have been shown to be a causative agent of preweaning diarrhea [16, 48, 49], a central nervous system disease of mink, termed shaking mink syndrome (SMS) has been earlier described in Denmark, Sweden, and Finland [21]. Intriguingly, a novel mink astrovirus was recently identified from brain homogenates of mink suffering from SMS and completely sequenced (SMS-AstV) [4]. Although HAstV-PS and SMS-AstV are phylogenetically related, whether cross-species transmission occurred between humans and minks still remains to be confirmed.

In the second report, a retrospective study identified astrovirus infection in three infants with severe combined immunodeficiency (SCID) after allogenic hematopoietic stem cell transplantation [75]. A combination of cell culture, electron microscopy, and sequence-independent single primer amplification unambiguously identified human astrovirus type 4 as the cause of infection in all three patients. One patient succumbed to the infection and upon autopsy, astrovirus was detected in the brain as well as other organs including bone marrow and intestinal tissue. This case report illustrates that astrovirus has the potential to spread systemically and cause severe disseminated lethal infection in highly immunocompromised pediatric patients. As previously noted [31, 45], it is an interesting observation that astrovirus infection appears to be more common than other viral agents of gastroenteritis (e.g., rotavirus and norovirus) in immune-compromised individuals, such as HIV-infected individuals, or patients who are undergoing tissue transplant or chemotherapy. While the reason for this is currently unknown, the fact that human astroviruses are an endemic pathogen worldwide may be a mitigating factor [31].

Innate Immunity

The first line of defense against invading pathogens is the innate immune response. Upon induction of a protective response against the pathogenic agent, the innate immune response will signal the adaptive and cellular immune systems to clear infection [53]. While observational and epidemiological data suggest that the adaptive immune response is the central mediator for protection against astrovirus infection in humans [45], studies from a small animal model of turkey astrovirus (TAstV) have provided convincing, direct evidence to suggest that innate immune responses are critical to controlling astrovirus infection in young animals [32, 33]. Regarding human astrovirus (HAstV), the evidence for a role in innate immunity in astrovirus infection comes from in vitro studies demonstrating an interaction between the HAstV capsid protein (CP) and complement factors resulting in inhibition of complement activation and downstream effector function [6, 28, 29, 35]. These interactions between the innate immune system and astroviruses are reviewed below and implications of these findings discussed.

Innate Immune Studies in Animals

Studies of the immune response to astrovirus infections in animals are limited as compared to those focused on molecular epidemiology and pathology. In most species, evidence of responding immune cells comes from histopathological observations of infected intestinal tissues. Reports of both bovine and ovine astrovirus infections describe a potential interaction between the immune system and astroviruses in the intestinal mucosa. In infected lambs, astrovirus particles were observed inside subepithelial macrophages [64]. Similarly, astrovirus infections in calves were associated with cells in and around the Peyer's patches [74]. Based on these limited reports the clinical or immunological significance of these observations is unclear. The observations could be that of local innate immune cells responding to infection; however, Woode et al. suggest that more studies are needed to rule out the possibility that the presence of astrovirus in these cells does not lead to a loss in normal function of the local immune response machinery increasing the animals' susceptibility to other pathogens [73, 74].

Interestingly, bovine astrovirus is not the only astrovirus to be implicated in immune suppression. The type-2 turkey astrovirus (TAstV-2), originally isolated as a contributing factor to a multifactorial disease syndrome of turkeys known as poult enteritis and mortality syndrome (PEMS), has also been suggested to have immunomodulatory properties. Studies by Qureshi et al. demonstrated lymphocyctes from TAstV-2-infected poults had a significant decrease in their response to mitogen, and postulated these changes may lead to significant immune dysfunction [59]. In subsequent studies, Qureshi et al. reported that TAstV-2 infections resulted in changes to innate immune cells as well [58]. Most notably, innate inflammatory cells appeared to have an impaired ability to migrate to the abdominal cavity in response to a sephadex-induced peritonitis, as compared with controls [58]. Functional examination of these responding cells demonstrated a reduced phagocytic activity, reduced bacterial killing activity, and reduced expression of the pro-inflammatory cytokines IL-1 and IL-6 [58]. Collectively, these results suggest that TAstV-2 infection suppresses elements of the innate immune system. Whether these effects are through direct interactions between the virus and innate immune cells, or through other factors is currently unclear. However, Koci et al. reported that turkey poults infected with TAstV-2 have a significant increase in the amount of active TGF-β in their serum [33].

In spite of the evidence that TAstV-2 may suppress the systemic immune response, the infection is still acute, so the question remains as to what host factors are involved in preventing persistent infections. Studies by both Qureshi et al. (2001) and Koci et al. (2004) report that macrophages and adherent splenocytes from TAstV-2-infected poults demonstrate an increase in their production of nitric oxide (NO) following ex vivo stimulation to LPS, as compared to control poults [32, 58]. Additional studies demonstrated that treating the avian macrophage cell line HD11 with TAstV-2 infectious particles induced production of NO in vitro through expression of inducible nitric oxide synthase (iNOS) [32]. This response was through a replication-independent mechanism [32] and can be induced using recombinant capsid protein (unpublished observation).

Collectively, these reports describe the innate immune factor produced by the host that could play a role in controlling astrovirus infections in the absence of an adaptive response. NO has long been recognized as a potent antiviral effector molecule; however, histological examination of TAstV-2 infected poults does not demonstrate large influxes of macrophages or other innate immune cells. The potential role NO may play in viral clearance was examined through a series of studies. Initially, the ability of NO and iNOS to affect viral replication was examined specifically. Virus was pretreated with NO donor compounds, which was found to inactivate the virus [32]. In the reciprocal experiment, embryonated turkey eggs were treated with specific iNOS inhibitors which resulted in a significant increase in virus replication [32]. Finally, tissues from TAstV-2-infected turkey poults were stained for the presence of TAstV-2 antigen and turkey iNOS. These experiments demonstrated that the virally infected tissues respond by upregulating the expression of iNOS. Most interestingly, the majority of the iNOS staining appears to be in the intestinal epithelial cells, suggesting that these cells are capable of mounting their own innate defense [46].

Innate Immune Studies in Humans

Very little is currently known about innate immune response to astrovirus in humans. Although differences exist between the TAstV animal model and human disease, the turkey animal model offers an excellent tool to examine astrovirus pathogenesis and immunity in order to better understand the role of innate immune responses in limiting viral replication in humans. In HAstV infections, the acute nature of the gastroenteritis, the short duration of symptoms, and the occurrence of most symptomatic infections in infants and young children who may lack adaptive immunity strongly suggest that innate responses may also play a key role in controlling and limiting disease. Thus, while preexisting antibodies may be key mediators of protection from reinfection, innate immune responses may be more relevant to control primary infections.

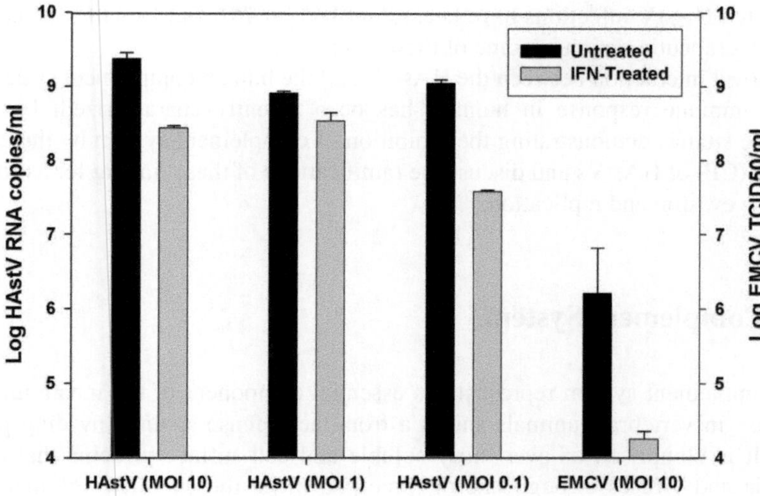

Fig. 5.1 Effect of IFN-α pretreatment on HAstV replication. Monolayers of CaCo-2 cells were pretreated with 1,000 U/ml IFN-α for 24 h or left untreated and then infected with HAstV-4 or encephalomyocarditis virus (EMCV) at the indicated multiplicities of infection (MOIs). Total virus was harvested at 24 h postinfection, and viral titers were determined. HAstV-4 was measured by quantitative RT-PCR and EMCV was measured by $TCID_{50}$ assay on Vero cells

In the past few decades, the knowledge of cell signaling pathways that lead to a type I interferon (IFN) response has increased tremendously, and it is now evident that many viruses encode strategies to subvert these cellular antiviral responses [68]. Infection of host cells by many RNA viruses results in the production of double-stranded RNA intermediates that are recognized by cellular Toll-like receptors or by the cytoplasmic RIG-I-like helicases, leading to the production of type I IFNs. Secreted IFN then acts in an autocrine and paracrine fashion to amplify this response and activate the expression of a variety of immunomodulatory and antiviral genes. Understanding the interplay between astroviruses and the innate antiviral responses, and determining if HAstV replication is sensitive to IFN would help to elucidate whether IFN could be a possible therapeutic option for treatment of persistent astrovirus gastroenteritis in specific cases. In vitro studies using Caco-2 cells have been performed in order to understand whether HAstV replication is sensitive to IFN. Caco-2 cells were pretreated with 1,000 U/ml of IFNα for 24 h before infection with HAstV-4 at different multiplicities of infection (MOIs), and virus yield was measured by quantitative RT-PCR. IFN-sensitive encephalomyocarditis virus (EMCV) was used as positive control [51]. As expected, IFN pretreatment significantly inhibited EMCV replication; however, HAstV-4 appears to be considerably less sensitive to IFN, especially at higher MOIs (Fig. 5.1) (Guix unpublished data). From a clinical standpoint, these preliminary data highlight the need for additional studies to further characterize the effect of IFN on HAstV replication. Although HAstV infection is generally considered a short-term, self-limiting illness, protracted HAstV infections have been described [8, 75], and IFN may be considered a therapeutic option in some of these cases.

A novel interaction between the HAstVs and the human complement system, an innate immune response in humans has been recently characterized. Here we describe studies demonstrating the inhibition of complement system by the capsid protein (CP) of HAstVs and discuss the ramifications of these finding for astrovirus immune evasion and replication.

The Complement System

The complement system represents an essential component of the innate immune response in vertebrate animals and is a frontline defense against invading pathogens. It is comprised of over thirty soluble and cell-surface proteins that detect bacteria and viruses through one of three pathways: the classical, the mannose-binding lectin, or the alternative pathways (Fig. 5.2) [66, 69]. Upon complement activation, pathogens are destroyed through a variety of mechanisms including the lysis of infected cells and certain pathogens, opsonization and phagocytosis, removal of pathogens via immune complexes, enhanced T and B cell priming as well as mediating a robust inflammatory response via leukocyte chemotaxis to the site of infection (Fig. 5.2) [69].

The classical pathway is initiated by C1, which is comprised of the pattern recognition molecule C1q and the serine–protease tetramer, C1s–C1r–C1r–C1s [9]. C1q consists of six stalks, or collagen-like regions (CLRs), attached to six globular-head regions (GHRs), forming a bouquet-like, quarternary structure [56]. The GHRs of C1q typically bind to clustered IgG or IgM initiating a conformational change in the CLRs [19]. The C1s–C1r–C1r–C1s complex resides within the N-terminal region of the CLRs and undergoes cleavage and activation when the CLRs initiate its conformational change [18]. This activation of C1 will then launch activation of the classical cascade pathway via C4 and then C2. C4 activation will generate the inflammatory peptide C4a and the opsonin C4b [24, 38].

The lectin pathway is initiated by the pattern recognition lectins in association with their cognate serine proteases, MASPs (MBL-associated serine proteases), which are homologous to the classical pathway serine protease C1s. The prototypical lectin is mannose/mannan binding lectin or MBL. MBL has a similar ultrastructure to C1q and typically binds to foreign sugars (i.e., mannose) initiating a conformational change [5]. The nested MASPs will be cleaved and activated in response to this conformational change initiating activation of the lectin pathway leading to subsequent activation of C4 and then C2 [70].

The alternative pathway does not have a pattern recognition component, but does undergo spontaneous activation which is referred to as tick-over. The alternative pathway is very important in dramatically amplifying specific activation of the classical or lectin pathways, creating a potent positive-feedback loop. Humans deficient in alternative pathway components (i.e., do not have a competent alternative

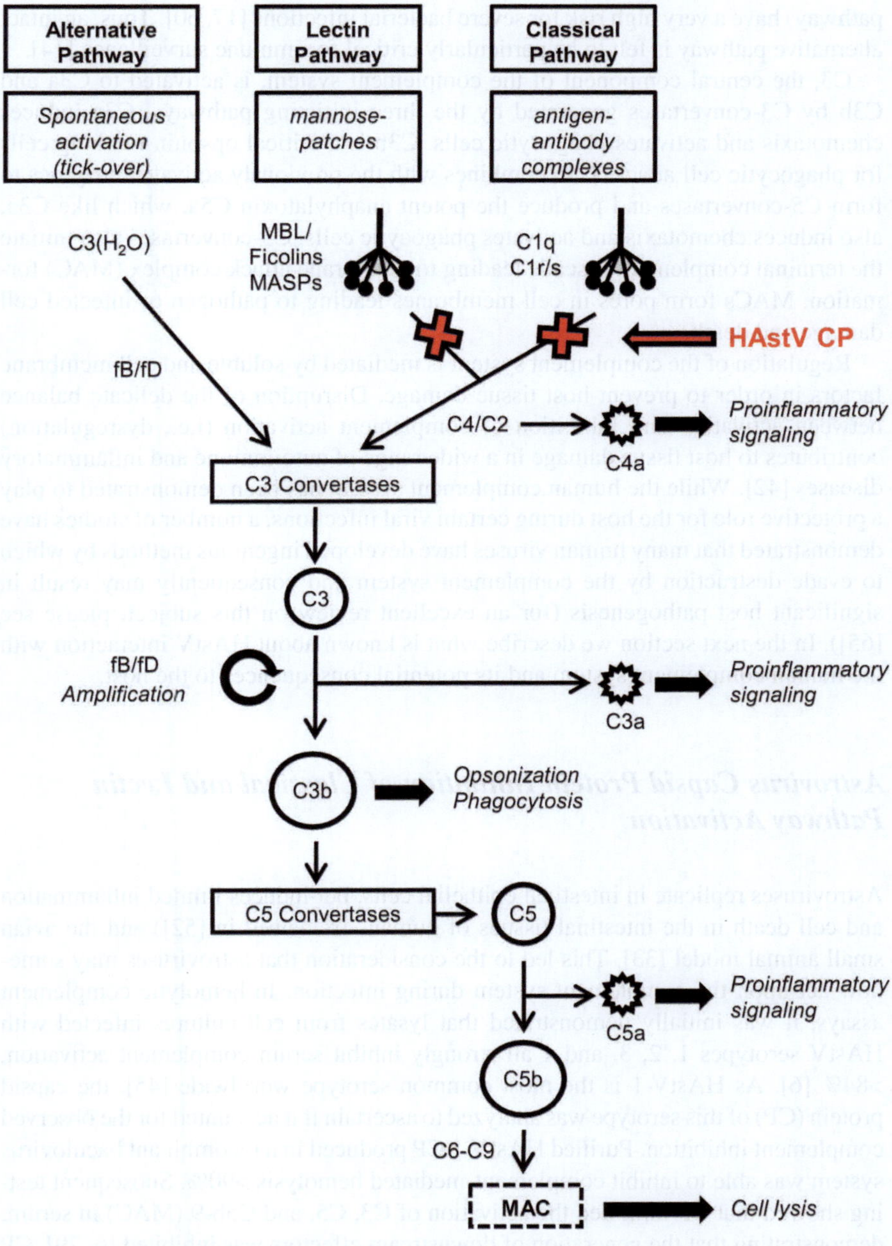

Fig. 5.2 Complement activation pathways and effectors. HAstV capsid protein (CP) blocks activation of the classical and lectin pathways

pathway) have a very high risk for severe bacterial infections [17, 60]. Thus, an intact alternative pathway is felt to be particularly critical for immune surveillance [14].

C3, the central component of the complement system, is activated to C3a and C3b by C3-convertases generated by the three initiating pathways. C3a induces chemotaxis and activates phagocytic cells. C3b is a critical opsonin marking cells for phagocytic cell attack. C3b combines with the previously activated enzymes to form C5-convertases and produce the potent anaphylatoxin C5a, which like C3a, also induces chemotaxis and activates phagocytic cells. C5-convertases also initiate the terminal complement cascade leading to membrane attack complex (MAC) formation. MACs form pores in cell membranes leading to pathogen or infected cell damage and death.

Regulation of the complement system is mediated by soluble and cell membrane factors in order to prevent host tissue damage. Disruption of the delicate balance between activation and inhibition of complement activation (i.e., dysregulation) contributes to host tissue damage in a wide range of autoimmune and inflammatory diseases [42]. While the human complement system has been demonstrated to play a protective role for the host during certain viral infections, a number of studies have demonstrated that many human viruses have developed ingenious methods by which to evade destruction by the complement system and consequently may result in significant host pathogenesis (for an excellent review on this subject, please see [65]). In the next section we describe what is known about HAstV interaction with the human complement system and its potential consequences to the host.

Astrovirus Capsid Protein Inhibition of Classical and Lectin Pathway Activation

Astroviruses replicate in intestinal epithelial cells, but induces limited inflammation and cell death in the intestinal tissues of humans (reviewed in [52]) and the avian small animal model [33]. This led to the consideration that astroviruses may somehow act upon the complement system during infection. In hemolytic complement assays, it was initially demonstrated that lysates from cell cultures infected with HAstV serotypes 1, 2, 3, and 4 all strongly inhibit serum complement activation, >84% [6]. As HAstV-1 is the most common serotype worldwide [45], the capsid protein (CP) of this serotype was analyzed to ascertain if it accounted for the observed complement inhibition. Purified HAstV-1 CP produced in a recombinant baculovirus system was able to inhibit complement-mediated hemolysis >90%. Subsequent testing showed that CP inhibited the activation of C3, C5, and C5b-9 (MAC) in serum, demonstrating that the generation of downstream effectors was inhibited [6, 29]. CP strongly inhibits activation of the classical pathway and antibody-initiated complement activation in a variety of systems, but has no significant effects on the alternative pathway. CP inhibited activation of C1, as assayed by C1s activation, preventing the activation of C4 in serum. Moreover, complement activation may be restored with the addition of exogenous C1. CP binds to both the globular head regions and

Table 5.2 Capsid protein peptides (CPP)

Peptide	Peptide sequence (N→C)
CPP2	NPVLVKDATGSTQFGPVQALGAQYSMWKLK
CPP1	PAICQRATATLGTVGSNTSGTTEIEACILL
8–22	PAICQRA---------------------------EIEACILL

collagen-like regions (CLR) of C1q, but has higher affinity for the CLR, where the C1s–C1r–C1r–C1s tetramer is nestled [29]. C1s–C1r–C1r–C1s is displaced from intact C1 by CP, suggesting the displacement of the cognate serine protease(s) is the likely mechanism of inhibiting C1 activation [29]. CP also inhibits mannan activation of MBL/MASP-2 in human serum [29]. MBL with a point mutation that prevents binding to MASP-2 (i.e., Lys55Gln substitution) cannot bind CP, suggesting that CP competes for the cognate serine protease binding site of MBL, similar to C1.

Astrovirus Capsid Protein Inhibition of Antibody-Initiated Complement Activation

Within the amino acid sequence of the CP molecule, a 60-residue region of homology corresponding to human neutrophil peptide 1 (HNP-1) was identified [28]. HNP-1 has been identified by other investigators as an inhibitor of C1 and MBL [26, 67]. From the region of homology, two synthetic peptides were generated of 30 residues each, coat protein peptide 1 (CPP1) and coat protein peptide 2 (CPP2) (Table 5.2). CPP1 inhibited the binding of CP to C1q and inhibited C1 activation, as assayed by C1s cleavage, suggesting a similar mechanism of action as the parent compound [28]. CPP2 did not show either of these properties, demonstrating that the active region of CP could be successfully narrowed to a small peptide construct. Additional truncations demonstrated that removing the central region of CPP1 (Δ8–22) yielded a functional peptide of 15 residues (Table 5.2) [28]. The Δ8–22 construct inhibited antibody-initiated complement activation to a greater degree than CPP1, as demonstrated in hemolytic as well as C4 activation assays and similar to the parent CP, the Δ8–22 construct still had minimal effect on the alternative pathway [28]. These efforts yielded a 15-residue peptide construct that is specific for inhibiting antibody-initiated complement activation. A model of the mechanism by which CP and the CPP derivatives inhibit C1 and MBL activation is presented in Fig. 5.3.

The sequence of CPP1 corresponds to HAstV-1 CP amino acid residues 79–108 located in the highly conserved domain of the CP. Clustal alignment of the complete CP sequence of 44 HAstV serotypes 1–8 demonstrates that the CPP1 sequence is highly conserved among the classical HAstVs as expected (Fig. 5.4, upper panel). Interestingly, the sequences from the recently described novel HAstVs are less conserved in this region (Fig. 5.4, lower panel). Analysis of the sequence alignment of HAstVs and the nonhuman AstVs shows that the CPP1 domain is very well conserved in the feline and porcine astrovirus isolates, but less so for the mink and

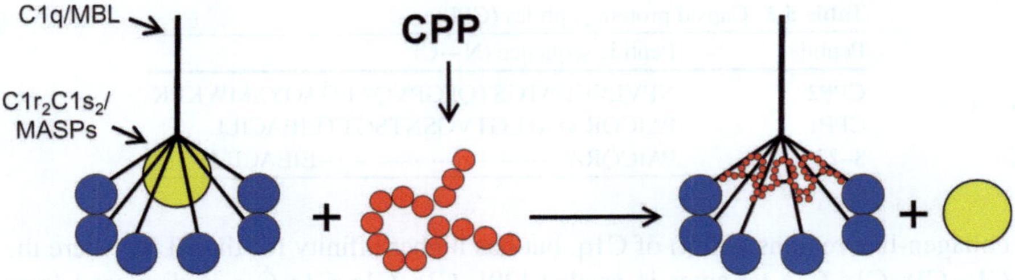

Fig. 5.3 A model of capsid protein peptide (CPP) interfering with the normal association of C1q/MBL with their cognate serine proteases (C1r2C1s2/MASPs). CPP functionally displaces the C1r2C1s2/MASPs by binding to the CLR of C1q/MBL, respectively

```
HAstV-1        (Accession #Z25771)     PAICQRATATLGTVGSNTSGTTEIEACILL
                                       ***** .******:*:**:*:****.****

HMOAstV-A      (Accession #GQ415660)   TKFTVKVSATIGKIGPNVNSGPELQISTFM
HAstV-VA2      (Accession #GQ502193)   TKFTVKVSATIGKIGPNVNSGPELQISTFM
HMOAstV-C      (Accession #GQ415662)   PRFAVTVSATIGKVGPNKSQGPELQISTFM
HMOAstV-VA1    (Accession #FJ973620)   PRFAVTVSATIGKVGPNKSQGPELQISTFM
HMOAstV-B      (Accession #GQ415661)   PRFSVSVSATIGKVGPNKAQGPELQISTFM
HAstV-MLB1     (Accession #FJ402983)   PAVSQTITATLGTVGSNLSDVVETECAVFL
HAstV-MLB2     (Accession #JF742759)   PAVSQTITATLGTVGSNLSDVVETECAVFL
                                       . .    :**:*.:*.*     * : . ::
```

Fig. 5.4 Alignment of HAstV CP sequences corresponding to CPP1. *Upper panel*: 44 HAstV serotypes 1–8 were aligned to the prototypical HAstV-1 CP residues 79–108 (CPP1 sequence) as determined by ClustalW2 analysis. *Asterisk* indicates identical residues, *colon* indicates conserved residues, *dot* indicates semi-conserved residues between all sequences. For clarity only HAstV-1 is shown. *Lower panel*: the recently discovered "nonclassical" HAstV CPs were aligned with the 44 HAstV CP sequences as described in the *upper panel*. For clarity, only seven novel HAstVs are shown. NCBI accession numbers for each isolate are given

ovine isolates. Avian astroviruses show poor homology. Whether the novel HAstVs, along with other nonhuman astroviruses also inhibit complement activation remains an interesting but unanswered question.

Implication of Astrovirus Capsid Protein Interactions with C1q and MBL

The interaction of the HAstV CP and the isolated peptides with C1q and MBL represents a novel mechanism by which astrovirus particles inhibit activation of the classical and lectin pathways of complement, respectively. However, in the absence of a small animal model to study HAstV pathogenesis, we can only speculate as to the function of C1q and MBL binding to the virus life cycle and pathogenesis. Many enteric pathogens such as rotavirus induce diarrhea by destroying enterocytes in the epithelium and inducing an increased inflammatory response [30]. However, astrovirus-induced diarrhea does not appear to result in significant cell death or inflammation in humans or in an avian astrovirus animal model (reviewed in [52]). One of the critical roles of complement is to initiate vigorous inflammatory (leukocyte) cell migration to infected tissues via the peptide mediators C3a/C5a which are generated by cleavage of C3 and C5, respectively [15]. Inhibition of iC3b, C5a, and SC5b-9 formation by CP reveals that CP inhibits formation of these chemoattractants and perhaps contributes to the limited leukocyte migration to the site of infection [6, 29]. Interestingly, C1 components (C1q, C1r, C1s) are synthesized mainly in columnar intestinal epithelial cells in vivo and are secreted into the lumen of the gastrointestinal tract (reviewed in [41]). Thus, the binding of C1q to CP may result in local complement suppression and aid in astrovirus immune evasion of the innate immune response.

In addition to immune evasion, there exists the possibility that astrovirus particles may bind C1q (or MBL) and utilize C1q receptors for cellular entry. Two well characterized, ubiquitously expressed cellular receptors for C1q have been described: gC1qR which binds the GHR of C1q and cC1qR (calreticulin) which binds the CLR of C1q (reviewed in [22]). Other receptors for C1q such as the integrin $\alpha 2\beta 1$ have been described as well [76]. Given the ability of astrovirus particles to bind C1q, it is conceivable that this event could enable virus coated with C1q to bind gC1qR/cC1qR/$\alpha 2\beta 1$ as receptors or co-receptors for cellular entry.

Conclusions

Although epidemiological studies have implicated the important role of antibody in astrovirus infection, and much progress has been made in elucidating the role of the innate immune response against human astrovirus as well as in a turkey astrovirus small animal model, we still have a very limited understanding of immune responses against human astrovirus. The development of a small animal model to study the immune response to human astrovirus would be a tremendous advancement and allow one to verify the role of the adaptive, cellular and innate immune responses that have been reported to date. A better understanding of the interplay of the immune system with human astroviruses is critical given the endemic nature of the pathogen, the rapidity at which new astroviruses are being discovered and the observations that astroviruses represent a significant disease burden among the young, elderly, and immune-compromised.

References

1. Bagci S, Eis-Hübinger AM, Franz AR, Bierbaum G, Heep A, Schildgen O, et al. Detection of astrovirus in premature infants with necrotizing enterocolitis. Pediatr Infect Dis J. 2008;27:347–50.
2. Bagci S, Eis-Hübinger AM, Yassin AF, Simon A, Bartmann P, Franz AR, et al. Clinical characteristics of viral intestinal infection in preterm and term neonates. Eur J Clin Microbiol Infect Dis. 2010;29:1079–84.

3. Björkholm M, Celsing F, Runarsson G, Waldenström J. Successful intravenous immunoglobulin therapy for severe and persistent astrovirus gastroenteritis after fludarabine treatment in a patient with Waldenström's macroglobulinemia. Int J Hematol. 1995;62:117–20.
4. Blomström AL, Widén F, Hammer AS, Belák S, Berg M. Detection of a novel astrovirus in brain tissue of mink suffering from shaking mink syndrome by use of viral metagenomics. J Clin Microbiol. 2010;48:4392–6.
5. Bohlson SS, Fraser DA, Tenner AJ. Complement proteins C1q and MBL are pattern recognition molecules that signal immediate and long-term protective immune functions. Mol Immunol. 2007;44:33–43.
6. Bonaparte RS, Hair PS, Banthia D, Marshall DM, Cunnion KM, Krishna NK. Human astrovirus coat protein inhibits serum complement activation via C1, the first component of the classical pathway. J Virol. 2008;82:817–27.
7. Burbelo PD, Ching KH, Esper F, Iadarola MJ, Delwart E, Lipkin WI, et al. Serological studies confirm the novel astrovirus HMOAstV-C as a highly prevalent human infectious agent. PLoS One. 2011;6:e22576. Epub 2011 Aug 4.
8. Caballero S, Guix S, El-Senousy WM, Calicó I, Pintó RM, Bosch A. Persistent gastroenteritis in children infected with astrovirus: association with serotype-3 strains. J Med Virol. 2003;71:245–50.
9. Cooper NR. The classical complement pathway: activation and regulation of the first complement component. Adv Immunol. 1985;37:151–216.
10. Coppo P, Scieux C, Ferchal F, Clauvel J, Lassoued K. Astrovirus enteritis in a chronic lymphocytic leukemia patient treated with fludarabine monophosphate. Ann Hematol. 2000;79:43–5.
11. Cox GJ, Matsui SM, Lo RS, Hinds M, Bowden RA, Hackman RC, et al. Etiology and outcome of diarrhea after marrow transplantation: a prospective study. Gastroenterology. 1994;107:1398–407.
12. Cubitt WD, Mitchell DK, Carter MJ, Willcocks MM, Holzel H. Application of electronmicroscopy, enzyme immunoassay, and RT-PCR to monitor an outbreak of astrovirus type 1 in a paediatric bone marrow transplant unit. J Med Virol. 1999;57:313–21.
13. Cunliffe NA, Glass RI. Gastrointestinal manifestations of HIV infection. Lancet. 1996;348:1037.
14. Daha MR. Role of complement in innate immunity and infections. Crit Rev Immunol. 2010;30:47–52. Review.
15. Ember JA, Jagels MA, Hugli TE. Characterization of complement anaphylatoxins and their biological responses. In: Volanakis JE, Frank MM, editors. The human complement system in health and disease. 1st ed. New York: Dekker; 1998.
16. Englund L, Chriél M, Dietz HH, Hedlund KO. Astrovirus epidemiologically linked to preweaning diarrhoea in mink. Vet Microbiol. 2002;85:1–11.
17. Figueroa JE, Densen P. Infectious diseases associated with complement deficiencies. Clin Microbiol Rev. 1991;4:359–95.
18. Gaboriaud C, Teillet F, Gregory LA, Thielens NM, Arlaud GJ. Assembly of C1 and the MBL- and ficolin-MASP complexes: structural insights. Immunobiology. 2007;212:279–88.
19. Gaboriaud C, Thielens NM, Gregory LA, Rossi V, Fontecilla-Camps JC, Arlaud GJ. Structure and activation of the C1 complex of complement: unraveling the puzzle. Trends Immunol. 2004;25:368–73.
20. Gallimore CI, Taylor C, Gennery AR, Cant AJ, Galloway A, Lewis D, et al. Use of a heminested reverse transcriptase PCR assay for detection of astrovirus in environmental swabs from an outbreak of gastroenteritis in a pediatric primary immunodeficiency unit. J Clin Microbiol. 2005;43:3890–4.
21. Gavier-Widén D, Broje C, Dietz HH, Englund L, Hammer AS, Hedlund KO, et al. Investigations into shaking mink syndrome: an encephalomyelitis of unknown cause in farmed mink (Mustela vison) kits in Scandinavia. J Vet Diagn Invest. 2004;16:305–12.
22. Ghebrehiwet B, Peerschke EI. cC1q-R (calreticulin) and gC1q-R/p33: ubiquitously expressed multi-ligand binding cellular proteins involved in inflammation and infection. Mol Immunol. 2004;41:173–83.

23. Giordano MO, Martinez LC, Rinaldi D, Espul C, Martinez N, Isa MB, et al. Diarrhea and enteric emerging viruses in HIV-infected patients. AIDS Res Hum Retroviruses. 1999;15:1427–32.
24. Gorski JP, Hugli TE, Müller-Eberhard HJ. Characterization of human C4a anaphylatoxin. J Biol Chem. 1981;256:2707–11.
25. Gray JJ, Wreghitt TG, Cubitt WD, Elliot PR. An outbreak of gastroenteritis in a home for the elderly associated with astrovirus type 1 and human calicivirus. J Med Virol. 1987;23:377–81.
26. Groeneveld TW, Ramwadhdoebé TH, Trouw LA, van den Ham DL, van der Borden V, Drijfhout JW, et al. Human neutrophil peptide-1 inhibits both the classical and the lectin pathway of complement activation. Mol Immunol. 2007;44:3608–14.
27. Grohmann GS, Glass RI, Pereira HG, Monroe SS, Hightower AW, Weber R, et al. Enteric viruses and diarrhea in HIV-infected patients. Enteric opportunistic infections working group. N Engl J Med. 1993;329:14–20.
28. Gronemus JQ, Hair PS, Crawford KB, Nyalwidhe JO, Cunnion KM, Krishna NK. Potent inhibition of the classical pathway of complement by a novel C1q-binding peptide derived from the human astrovirus coat protein. Mol Immunol. 2010;48:305–13.
29. Hair PS, Gronemus JQ, Crawford KB, Salvi VP, Cunnion KM, Thielens NM, et al. Human astrovirus coat protein binds C1q and MBL and inhibits the classical and lectin pathways of complement activation. Mol Immunol. 2010;47:792–8.
30. Kapikian AZ, Hoshino Y, Chanock RM. Rotaviruses. In: Knipe DM, Howley PM, editors. Fields virology, vol. 2. 4th ed. Philadelphia, PA: Lippincott, Williams & Wilkins; 2001.
31. Koci MD. Immunity and resistance to astrovirus infection. Viral Immunol. 2005;18:11–6. Review.
32. Koci MD, Kelley LA, Larsen D, Schultz-Cherry S. Astrovirus-induced synthesis of nitric oxide contributes to virus control during infection. J Virol. 2004;78:1564–74.
33. Koci MD, Moser LA, Kelley LA, Larsen D, Brown CC, Schultz-Cherry S. Astrovirus induces diarrhea in the absence of inflammation and cell death. J Virol. 2003;77:11798–808.
34. Koopmans MP, Bijen MH, Monroe SS, Vinjé J. Age-stratified seroprevalence of neutralizing antibodies to astrovirus types 1 to 7 in humans in The Netherlands. Clin Diagn Lab Immunol. 1998;5:33–7.
35. Krishna NK, Cunnion KM. Human astrovirus coat protein: a novel C1 inhibitor. Adv Exp Med Biol. 2008;632:237–51. Review.
36. Kurtz JB, Lee TW. Astrovirus gastroenteritis age distribution of antibody. Med Microbiol Immunol. 1978;166:227–30.
37. Kurtz JB, Lee TW, Craig JW, Reed SE. Astrovirus infection in volunteers. J Med Virol. 1979;3:221–30.
38. Law SK, Dodds AW. The internal thioester and the covalent binding properties of the complement proteins C3 and C4. Protein Sci. 1997;6:263–74. Review.
39. Lewis DC, Lightfoot NF, Cubitt WD, Wilson SA. Outbreaks of astrovirus type 1 and rotavirus gastroenteritis in a geriatric in-patient population. J Hosp Infect. 1989;14:9–14.
40. Liste MB, Natera I, Suarez JA, Pujol FH, Liprandi F, Ludert JE. Enteric virus infections and diarrhea in healthy and human immunodeficiency virus-infected children. J Clin Microbiol. 2000;38:2873–7.
41. Loos M. Biosynthesis of the collagen-like C1q molecule and its receptor functions for Fc and polyanionic molecules on macrophages. Curr Top Microbiol Immunol. 1983;102:1–56.
42. Markiewski MM, Lambris JD. The role of complement in inflammatory diseases from behind the scenes into the spotlight. Am J Pathol. 2007;171:715–27. Review.
43. Marshall JA, Bruggink LD, Sturge K, Subasinghe N, Tan A, Hogg GG. Molecular features of astrovirus associated with a gastroenteritis outbreak in an aged-care centre. Eur J Clin Microbiol Infect Dis. 2007;26:67–71.
44. Matsui SM, Lewis TL, Chiu E, Smith LS, Dupis K, Cahill CK, et al. A 27-nm virus isolated during an outbreak of acute infectious nonbacterial gastroenteritis in a convalescent hospital: a possible new serotype. J Infect Dis. 1981;143:791–5.
45. Mendez E, Arias CF. Astroviruses. In: Knipe DM, Howley PM, editors. Fields virology, vol. 1. 5th ed. Philadelphia, PA: Lippincott, Williams & Wilkins; 2007.

46. Meyerhoff RR, Nighot PK, Ali RA, Blikslager AT, Koci MD. Characterization of turkey inducible nitric oxide synthase and identification of its expression in the intestinal epithelium following astrovirus infection. Comp Immunol Microbiol Infect Dis. 2012;35:63–9.
47. Mitchell DK. Astrovirus gastroenteritis. Pediatr Infect Dis J. 2002;21:1067–9. Review.
48. Mittelholzer C, Englund L, Hedlund KO, Dietz HH, Svensson L. Detection and sequence analysis of Danish and Swedish strains of mink astrovirus. J Clin Microbiol. 2003;41:5192–4.
49. Mittelholzer C, Hedlund KO, Englund L, Dietz HH, Svensson L. Molecular characterization of a novel astrovirus associated with disease in mink. J Gen Virol. 2003;84:3087–94.
50. Molberg O, Nilsen EM, Sollid LM, Scott H, Brandtzaeg P, Thorsby E, et al. CD4+ T cells with specific reactivity against astrovirus isolated from normal human small intestine. Gastroenterology. 1998;114:115–22.
51. Morrison JM, Racaniello VR. Proteinase 2Apro is essential for enterovirus replication in type I interferon-treated cells. J Virol. 2009;83:4412–22.
52. Moser LA, Schultz-Cherry S. Pathogenesis of astrovirus infection. Viral Immunol. 2005;18: 4–10. Review.
53. Murphy KP. Innate immunity: the first line of defense. In: Janeway CA, editor. Immunobiology. 8th ed. London: Garland Science; 2012.
54. Noel J, Cubitt D. Identification of astrovirus serotypes from children treated at the hospitals for sick children, London 1981-93. Epidemiol Infect. 1994;113:153–9.
55. Oshiro LS, Haley CE, Roberto RR, Riggs JL, Croughan M, Greenberg H, et al. A 27-nm virus isolated during an outbreak of acute infectious nonbacterial gastroenteritis in a convalescent hospital: a possible new serotype. J Infect Dis. 1981;143:791–5.
56. Poon PH, Schumaker VN, Phillips ML, Strang CJ. Conformation and restricted segmental flexibility of C1, the first component of human complement. J Mol Biol. 1983;168:563–77.
57. Quan PL, Wagner TA, Briese T, Torgerson TR, Hornig M, Tashmukhamedova A, et al. Astrovirus encephalitis in boy with X-linked agammaglobulinemia. Emerg Infect Dis. 2010;16:918–25.
58. Qureshi MA, Saif YM, Heggen-Peay CL, Edens FW, Havenstein GB. Induction of functional defects in macrophages by a poult enteritis and mortality syndrome-associated turkey astrovirus. Avian Dis. 2001;45:853–61.
59. Qureshi MA, Yu M, Saif YM. A novel "small round virus" inducing poult enteritis and mortality syndrome and associated immune alterations. Avian Dis. 2000;44:275–83.
60. Ram S, Lewis LA, Rice PA. Infections of people with complement deficiencies and patients who have undergone splenectomy. Clin Microbiol Rev. 2010;23:740–80. Review.
61. Sebire NJ, Malone M, Shah N, Anderson G, Gaspar HB, Cubitt WD. Pathology of astrovirus associated diarrhoea in a paediatric bone marrow transplant recipient. J Clin Pathol. 2004;57:1001–3.
62. Shastri S, Doane AM, Gonzales J, Upadhyayula U, Bass DM. Prevalence of astroviruses in a children's hospital. J Clin Microbiol. 1998;36:2571–4.
63. Sirinavin S, Techasaensiri C, Okascharoen C, Nuntnarumit P, Tonsuttakul S, Pongsuwan Y. Neonatal astrovirus gastroenteritis during an inborn nursery outbreak. J Hosp Infect. 2006;64:196–7.
64. Snodgrass DR, Angus KW, Gray EW, Menzies JD, Paul G. Pathogenesis of diarrhoea caused by astrovirus infections in lambs. Arch Virol. 1979;60:217–26.
65. Stoermer KA, Morrison TE. Complement and viral pathogenesis. Virology. 2011;41:362–73.
66. Thiel S, Vorup-Jensen T, Stover CM, Schwaeble W, Laursen SB, Poulsen K, et al. A second serine protease associated with mannan-binding lectin that activates complement. Nature. 1997;386:506–10.
67. van den Berg RH, Faber-Krol MC, van Wetering S, Hiemstra PS, Daha MR. Inhibition of activation of the classical pathway of complement by human neutrophil defensins. Blood. 1998;92:3898–903.
68. Versteeg GA, García-Sastre A. Viral tricks to grid-lock the type I interferon system. Curr Opin Microbiol. 2010;13:508–16.

69. Volanakis JE. Overview of the complement system. In: Volanakis JE, Frank MM, editors. The human complement system in health and disease. 1st ed. New York: Dekker; 1998.
70. Wallis R, Mitchell DA, Schmid R, Schwaeble WJ, Keeble AH. Paths reunited: Initiation of the classical and lectin pathways of complement activation. Immunobiology. 2010;215:1–11.
71. Watkins JS. Astrovirus gastroenteritis on a geriatric ward. Commun Dis Rep. 1984;84:3.
72. Wood DJ, David TJ, Chrystie IL, Totterdell B. Chronic enteric virus infection in two T-cell immunodeficient children. J Med Virol. 1998;24:435–44.
73. Woode GN, Gourley NE, Pohlenz JF, Liebler EM, Mathews SL, Hutchinson MP. Serotypes of bovine astrovirus. J Clin Microbiol. 1985;22:668–70.
74. Woode GN, Pohlenz JF, Gourley NE, Fagerland JA. Astrovirus and Breda virus infections of dome cell epithelium of bovine ileum. J Clin Microbiol. 1984;19:623–30.
75. Wunderli W, Meerbach A, Guengoer T, Berger C, Greiner O, Caduff R, et al. Astrovirus infection in hospitalized infants with severe combined immunodeficiency after allogeneic hematopoietic stem cell transplantation. PLoS One. 2011;6:e27483. Epub 2011 Nov 11.
76. Zutter MM, Edelson BT. The alpha2beta1 integrin: a novel collectin/C1q receptor. Immunobiology. 2007;212:343–53.

Chapter 6
Astrovirus Taxonomy

Susana Guix, Albert Bosch, and Rosa M. Pintó

Abstract Since their first discovery in children, astroviruses have been isolated from stools from a wide variety of other mammals where they have been investigated (cats, cattle, deer, dogs, mice, pigs, sheep, minks, bats, cheetahs, rats, rabbits, and marine mammals), as well as birds (turkeys, chickens, ducks, guinea fowls, and pigeons). They are nonenveloped positive-sense, single-stranded RNA viruses classified as the *Astroviridae* family.

The family is divided into two genera: genus *Mamastrovirus* (*MAstVs*) including viruses infecting mammals, and genus *Avastrovirus* (*AAstVs*) including viruses infecting avian species. Based on the species of the host of origin, the Ninth ICTV Report described six virus species within the *MAstV* genus and three species within the *AAstV* genus. However, the recent discovery of many novel mammalian and avian astroviruses highlighted the fact that astrovirus species cannot be established based solely on the animal species of the host, and the *Astroviridae* Study Group soon updated the taxonomic scheme. Demarcation into genotype species within genera is now performed on the basis of host range but also genetic differences on the complete capsid sequence.

Following these standardized criteria, the *MAstV* and the *AAstV* genera may be divided into 33 and 7 species, respectively, showing that astroviruses constitute a remarkably genetically diverse family with a broad spectrum of host species. Members of multiple astrovirus species can infect the same host, while other variants may infect more than one animal species. Since new astroviruses are continuously being discovered, their taxonomy should be regularly updated using the current classification scheme and criteria, which are summarized in this chapter.

S. Guix, Ph.D. (✉) • A. Bosch, Ph.D. • R.M. Pintó, Ph.D.
Department of Microbiology, Enteric Virus Laboratory, Institute of Nutrition and Food Safety, University of Barcelona, Avenue Diagonal 643, Barcelona 08028, Spain
e-mail: susanaguix@ub.edu; abosch@ub.edu; rpinto@ub.edu

S. Schultz-Cherry (ed.), *Astrovirus Research: Essential Ideas, Everyday Impacts, Future Directions*, DOI 10.1007/978-1-4614-4735-1_6,

Historic Perspective and Present Status of Astrovirus Taxonomy

Astroviruses were originally identified from the feces of infants with gastroenteritis and their name was derived from their distinctive appearance in electron microscopy—a smooth or slightly indented outer electron-dense shell with an inner, negatively staining five or six pointed star-shaped core [38]. They were included within the group of "small round-structured viruses" (SRSV) together with noroviruses and other gastroenteritis viruses. Since their first discovery in children, astroviruses have been isolated from stools from a wide variety of other mammals where they have been investigated (cats, cattle, deer, dogs, mice, pigs, sheep, mink, bats, cheetahs, rats, rabbits, and marine mammals), as well as birds (turkeys, chickens, ducks, guinea fowls, and pigeons) [11, 30, 33, 41, 50, 77, 78]. While in mammals astroviruses are mainly associated with diarrhea in young individuals or asymptomatic infections, in birds they cause a wider spectrum of diseases, including diarrhea, hepatitis, and nephritis. However, extraintestinal infections are not restricted to avian astroviruses. Recently, mink astroviruses have also been isolated from brain tissue of individuals suffering from the shaking mink syndrome (SMS) [3], and human astroviruses (HAstV) have been found in the brain of a 15-year-old boy with X-linked agammaglobulinemia suffering encephalitis [51]. Finally, human astroviruses have been also isolated from plasma of a febrile child [22], highlighting again the fact that astroviruses may go systemic and have effects outside the enteric system.

The family *Astroviridae* was permanently established within the positive-sense RNA virus *subphylum* in 1995, following the confirmation of several features of their molecular biology. In 1991, comparison of the proteins and RNAs synthesized in astrovirus-infected cells with those of other well-recognized families of nonenveloped +ssRNA animal viruses deterred the classification of astroviruses as members of either *Caliciviridae* or *Picornaviridae* families, as it had been initially proposed [47]. Later on, the sequence of the subgenomic RNA was determined, supporting the establishment of the *Astroviridae* family, as a new family of +ssRNA viruses [46]. At the same time, the analysis of the first complete sequence of a human astrovirus [26] identified three open reading frames (ORFs), ORF1a (encoding the protease motif), ORF1b (encoding the RNA polymerase), and ORF2 (encoding the capsid proteins). The same work pointed out two important findings that distinguish astroviruses from other nonenveloped +ssRNA viruses: the lack of a RNA-helicase domain, and the presence of a ribosomal frameshifting signal (RFS) between ORF1a and ORF1b. In 1994, the viral RNA-dependent RNA polymerase (RdRp) motif was characterized [36], and it was in 1995 when the International Committee on the Taxonomy of Viruses (ICTV) classified astroviruses as a new separate family named *Astroviridae* in the Sixth ICTV Report [45]. In summary, the main characteristics of the *Astroviridae* family are highlighted in Table 6.1. Among these, the three virological features which, altogether, distinguish astroviruses from other +ssRNA viruses are: the location of ORF1a and ORF1b encoding for nonstructural proteins at the 5′ end of the genome and ORF2 encoding the structural protein at the 3′ end; the use of a RFS mechanism to translate the RdRp encoded in ORF1b; and the lack of a helicase domain, characteristic of +ssRNA viruses with

Table 6.1 Main characteristics of the *Astroviridae* family

Virion structure	
	Icosahedric particles of 28–41 nm in diameter
	Nonenveloped
	Distinctive five- or six-pointed star shape under the electron microscope in about 10% of virions
Genome	
	One molecule of infectious, positive-sense, single-stranded RNA ranging between 6.4 and 7.7 kb in size, with a poly(A) tail
	ORFs encoding the nonstructural proteins (ORF1a and ORF1b) at the 5′-end, and ORF encoding the structural proteins (ORF2) at the 3′-end
	VPg protein linked to the 5′-end
	Presence of a ribosomal frameshifting signal (RFS) between ORF1a and ORF1b
	Lack of helicase domain
Replication	
	Use of a ~2.8 kb subgenomic RNA for the synthesis of structural proteins

genomes larger than 6 kb. The presence of a VPg protein linked to the 5'-end of the astrovirus genome has been speculated for many years by several authors, and its existence has been recently confirmed for human astroviruses [19].

Initially, the *Astroviridae* family contained only one genus, *Astrovirus* [45], and the classification was based on virus morphology, but later on the ICTV proposed a new system for classification and nomenclature, based on the host of origin. The family was divided into two genera: genus *Mamastrovirus* (*MAstVs*) including viruses infecting mammals, and genus *Avastrovirus* (*AAstVs*) including viruses infecting avian species. Within genera, astroviruses were further subdivided into virus species based on the animal species of the host from which the virus was isolated. Since then, this classification has not changed, and the current Ninth ICTV Report describes six species within the *Mamastrovirus* genus (*Bovine astrovirus, Feline astrovirus, Human astrovirus, Mink astrovirus, Ovine astrovirus*, and *Porcine astrovirus*), and three species within the *Avastrovirus* genus (*Chicken astrovirus, Duck astrovirus*, and *Turkey astrovirus*) [6]. Within species, viruses are classified in serotypes on the basis of 20-fold or greater, two-way cross-neutralization titers [6]. However, since most astroviruses are difficult to grow in cell culture and/or animal models and there is a strong correlation between serotypes and genotypes, serological classification has traditionally been speculated based on percentage of similarity of the capsid sequence.

The scenario of the astrovirus field has changed drastically since 2008. The use of viral metagenomic tools for large-scale surveillance studies and the analysis of the fecal virome of selected animal species have led to the identification of many novel astroviruses. These new data denoted that the current ICTV classification criteria required being revised and updated. Phylogenetic characterization of some of these viruses has shown that a single host species may be susceptible to infection by divergent astroviruses. For instance, this is the case of the recently described new human astroviruses, namely HAstV-MLB1 and MLB-2; HMOAstVs A, B, and C,and HAstV-VA1 and VA2, which have been found in patients suffering from gastroenteritis but are genetically unrelated to the "classical" serotypes infecting humans [13, 14, 15, 16, 17, 31]. Indeed, some of these new strains are even closer to some other animal astroviruses such as mink, sheep, or rat astroviruses, rather than to the classical human viruses, showing that astrovirus species proposed in the Ninth ICTV Report do not correlate with genetic lineages [9, 31]. A similar observation has been reported for astroviruses infecting pigs [37], bats [10, 79], California sea lions [55], sheep [53], mink [3], and turkeys [49]. These groups of viruses show a remarkably high level of genetic diversity, and viruses isolated from the same animal species cluster in different groups following phylogenetic reconstruction. In addition, some of them even cause different diseases.

On the other hand, it has also been observed that some astroviruses are not species specific and can infect more than one animal species. Among other examples, this would be the case of some bat astroviruses (BatAstV), which can infect different species of insectivorous bats [10, 79], the astrovirus identified in cheetahs, which presumably is the same virus that infects cats [1], or the Avian nephritis virus, which can infect both chickens and turkeys [23].

For all these reasons, all members of the *Astroviridae* Study Group submitted in 2010 two new taxonomic proposals to redefine the astrovirus classification criteria. According to the proposals, demarcation into species is performed not only on the basis of host range but also on phylogenetic differences [4, 5]. Within each genus, astroviruses are classified into genotype species (in italics), based on both genetic analysis of the full-length ORF2 encoding the capsid proteins, and the host of origin. Initially, the proposals suggested the use of capital letters to name the different genotypes, and grouped genotypes in larger genetic clusters or genogroups, which would be named using Roman numerals (GI.A, GI.B, etc., and GII.A, GII.B, etc.). However, following the ICTV recommendations, the proposals were modified. Genogroups were removed, and genotypes within each genus were named using Arabic numbers. Since the initial proposals were posted on the ICTV Web site for a while, the first nomenclature has been used in several publications. For the sake of clarity, tables presented in this chapter will include both nomenclatures (Tables 6.2 and 6.3).

This chapter revises the novel system concerning astrovirus taxonomy proposed and supervised by the *Astroviridae* Study Group [4, 5] soon after the publication of the Ninth ICTV Report [6]. It focuses on a comprehensive analysis of astrovirus variability including the most recent mammal and avian astrovirus isolates which were not included in the proposals [4, 5]. The discovery and characterization of novel viruses is important for the understanding of many human and animal diseases, and may lead to the identification of newly emerging pathogens in the future.

Astrovirus Infecting Mammals (Genus *Mamastrovirus*)

So far, astroviruses have been isolated from 31 species of mammals, 14 of them corresponding to different species of bats, which are thought to act as a reservoir [10, 75, 79]. Using the criteria suggested in the taxonomic proposal [4], mamastroviruses

Table 6.2 List of genotype species within the genus *Mamastrovirus* (*MAstV*)

Genotype species[a]	Host species of origin	Representative strain(s)	Alternative names temporarily reported in ICTV Web site	Serotypes/ lineages	Accession number of representative strain(s)	Amino acid p-dist within genotype	Associated disease(s)	Reference(s)
MAstV 1[b]	Human (*Homo sapiens*)	Human astrovirus 1 (HAstV-1)	GI.A	HAstV-1	L23513; Z25771; AY720892	0.276	Gastroenteritis	[43]
				HAstV-2	L13745			
				HAstV-3	AF141381			
				HAstV-4	Z33883; AY720891; DQ070852; DQ344027			
				HAstV-5	U15136; DQ028633			
				HAstV-6	Z46658; GQ495608			
				HAstV-7	Y08632			
				HAstV-8	AF260508			
MAstV 2[b]	Cat (*Felis catus*)	Feline astrovirus (FAstV)	GI.B		AF056197	–	Diarrhea in kittens	[23]
	Cheetah (*Acinonyx jubatus*)				EU650331[d]		Lethargy, anorexia, diarrhea, and regurgitation in young and adult animals (cheetah)	[1]
MAstV 3[b]	Pig (*Sus scrofa*)	Porcine astrovirus 1 (PAstV 1)	GI.C		AB037272; Y15938; HM756258	0.147	Gastroenteritis mainly in young piglets; asymptomatic infections	[27, 28, 37]

(continued)

Table 6.2 (continued)

Genotype species[a]	Host species of origin	Representative strain(s)	Alternative names temporarily reported in ICTV Web site	Serotypes/ lineages	Accession number of representative strain(s)	Amino acid p-dist within genotype	Associated disease(s)	Reference(s)
MAstV 4	California sea lion (*Zalophus californianus*)	California sea lion astrovirus 2 (CslAstV 2)	GI.D		FJ890352		Asymptomatic infections	[55]
MAstV 5	Dog (*Canis lupus familiaris*)	Canine astrovirus (CaAstV)	GI.E		FM213330; FM213331; FM213332	0.157	Gastroenteritis in puppies	[68]
					HM045005			[40]
					JN193534			[39]
					HQ623147; HQ623148; GU376736			[80]
MAstV 6	Human (*Homo sapiens*)	Human astrovirus MLB 1 (HAstV-MLB 1)	GI.F	MLB-1	FJ222451; FJ402983	0.181	Diarrhea; asymptomatic infections	[14]
				MLB-2	JF742759		Diarrhea; asymptomatic infections; viremia in a febrile child	[22]
MAstV 7	Bottlenose dolphin (*Tursiops truncates*)	Bottlenose dolphin astrovirus (BdAstV)	GI.G		FJ890355	–	Asymptomatic infections	[55]
MAstV 8	Human (*Homo sapiens*)	Human astrovirus HMO-A (HAstV-HMO-A)	GII.A		GQ415660	0.006	Diarrhea; detected also in children with nonpolio-virus acute flaccid paralysis	[31]
		Human astrovirus VA2 (HAstV-VA2)			GQ502193		Diarrhea	[17]

MAstV 9	Human (*Homo sapiens*)	Human astrovirus HMO-B (HAstV-HMO-B)	GII.B		GQ415661	0.138	Diarrhea; detected also in children with nonpolio-virus acute flaccid paralysis	[31]
		Human astrovirus HMO-C (HAstV-HMO-C)			GQ415662		Diarrhea; detected also in children with nonpolio-virus acute flaccid paralysis	[31]
		Human astrovirus VA1 (HAstV-VA1)			FJ973620		Diarrhea; Infection related to the onset of celiac disease in one patient	[16, 60]
		Human astrovirus PS (HAstV-PS)			GQ891990		Encephalitis in a young boy with X-agammaglobulinemia	[51]
MAstV 10[b]	Mink (*Neovison vison*)	Mink astrovirus (MAstV)	GII.C		AY179509	–	Preweaning diarrhea syndrome	[44]
MAstV 11	California sea lion (*Zalophus californianus*)	California sea lion AstV 1 (CslAstV 1)	GII.D		FJ890351	–	Asymptomatic infections	[55]
MAstV 12	Bat (*Taphozous melanopogon*)	TmAstV/GX/LD71/07 (group 3[c])	GII.E		FJ571067	–	Apparently asymptomatic infections	[79]
MAstV 13[b]	Sheep (*Ovis aries*)	Ovine astrovirus 1 (OAstV 1)	GII.F		NC_002469	–	Mild diarrhea in 2-day-old gnotobiotic lambs	[28, 29, 62]
MAstV 14	Bat (*Miniopterus magnate* and other bat species)	MmAstV/HK/AFCD57/05 (group 2[c])	GII.G		EU847144	–	Apparently asymptomatic infections	[10, 79]
MAstV 15	Bat (*Taphozous melanopogon* and other bat species)	TmAstV/GX/LD77/07 (group 2[c])	GII.H		FJ571066	–	Apparently asymptomatic infections	[79]

(continued)

Table 6.2 (continued)

Genotype species[a]	Host species of origin	Representative strain(s)	Alternative names temporarily reported in ICTV Web site	Serotypes/lineages	Accession number of representative strain(s)	Amino acid p-dist within genotype	Associated disease(s)	Reference(s)
MAstV 16	Bat (Pipistrellus abramus and other bat species)	PaAstV/HK/AFCD11/05 (group 2[c])	GII.I		EU847145	–	Apparently asymptomatic infections	[10, 79]
MAstV 17	Bat (*Hipposideros pomona, Hipposideros armiger* and other bat species)	HpAstV/GX/LC03/07 (group 7[c])	GII.J		FJ571074	0.318	Apparently asymptomatic infections	[79]
		HaAstV/GX/LS11/07 (group 7[c])			FJ571068			
MAstV 18	Bat (*Miniopterus pusillus* and other bat species)	MpAstV/HK/AFCD337/06 (group 5[c])	GII.K		EU847155	–	Apparently asymptomatic infections	[10, 79]
MAstV 19	Bat (*Taphozous melanopogon*)	TmAstV/GX/LD38/07 (group 6[c])	GII.L		FJ571065	0.184	Apparently asymptomatic infections	[79]
		TmAstV/GX/LD04/07 (group 6[c])			FJ571069			
		TmAstV/GX/LD27/07 (group 6[c])			FJ571070			
		TmAstV/GX/DX19/07 (group 6[c])			FJ571071			
		TmAstV/GX/LD45/07 (group 6[c])			FJ571072			
		TmAstV/GX/LD54/07 (group 6[c])			FJ571073			
MAstV 20	Mouse (*Mus musculus*)	Mouse astrovirus (Mouse AstV)	–		JF755422	–	–	[50]
MAstV 21	Mink (*Neovison vison*)	Mink astrovirus (SMS AstV)	–		GU985458	–	Shaking mink syndrome (SMS)	[3]

MAstV 22	Pig (*Sus scrofa*)	Porcine astrovirus 3 (PastV 3)	–	HM756261	–	Asymptomatic in suckling piglets	[37]
MAstV 23	Rabbit (*Oryctolagus cuniculus*)	Rabbit astrovirus (RabAstV)	–	JN052023	–	Enterocolitis or enteritis	[41]
MAstV 24	Pig (*Sus scrofa*)	Porcine astrovirus 5 (PAstV 5)	–	JF713711	0.223	–	[57]
				JN088537		–	[35]
	Sheep (*Ovis aries*)	Ovine astrovirus 2 (OAstV 2)		JN592482		Asymptomatic in healthy young animals (less than 3 weeks of age)	[53]
MAstV 25	Brown rats (*Rattus norvegicus*)	Rat astrovirus (RatAstV)	–	HM450381; HM450382	0.273	–	[9]
MAstV 26	Pig (*Sus scrofa*)	Porcine astrovirus 4 (PAstV 4) (PAstV/Hungary/2009)	–	GU562296	–	Asymptomatic infections	[52]
MAstV 27	Pig (*Sus scrofa*)	Porcine astrovirus 4 (PAstV 4-35-USA)	–	JF713713	–	–	[57]
MAstV 28	Cattle (*Bos taurus*)	Bovine astrovirus (BAstV34/HK)	–	HQ916315	–	Asymptomatic in adult cattle	[69]
MAstV 29	Cattle (*Bos taurus*)	Bovine astrovirus (BAstV76/HK)	–	HQ916316	–	Asymptomatic in adult cattle	[69]
MAstV 30	Cattle (*Bos taurus*)	Bovine astrovirus (BAstV170/HK)	–	HQ916314	–	Asymptomatic in adult cattle	[69]
MAstV 31	Pig (*Sus scrofa*)	Porcine astrovirus 2 (PAstV 2/14-4/Canada/2)	–	HM756260	–	Asymptomatic infections	[37]
MAstV 32	Pig (*Sus scrofa*)	Porcine astrovirus 2 (PAstV 2/12-4/Canada/2)	–	HM756259	0.301	Asymptomatic infections	[37]
				HQ647383		Diarrhea	[34]
				JF713710; JF713712			[57]

(continued)

Table 6.2 (continued)

Genotype species[a]	Host species of origin	Representative strain(s)	Alternative names temporarily reported in ICTV Web site	Serotypes/ lineages	Accession number of representative strain(s)	Amino acid p-dist within genotype	Associated disease(s)	Reference(s)
MAstV 33	Cattle (*Bos taurus*)	Bovine astrovirus (BAstV18/HK)	–		HQ916317; HQ916313	0.268	Asymptomatic in adult cattle	[69]
	Roe deer (*Capreolus capreolus*)	*Capreolus capreolus* astrovirus 1 (CcAstV 1)		CcAstV-1 CcAstV-2	HM447045 HM447046		Diarrhea	[61]

[a]Genotype species are in italics. Genotypes in italics and bold are included in the Taxonomic Proposal submitted to the ICTV by the *Astroviridae* Study Group in 2010 [4]. Genotypes in italics but regular type are proposed in this chapter, following the same criteria

[b]Astrovirus species considered in the Ninth ICTV Report [6]

[c]Monophyletic groups based on analysis of a partial sequence of ORF1b gene [78]

[d]Partial ORF2 sequences available only

Table 6.3 List of genotype species within the genus *Avastrovirus* (*AAstV*)

Geno type species[a]	Host species of origin	Representative strain(s)	Other names temporarily reported in ICTV Web site	Serotypes/ lineages	Accession number of representative strain(s)	Amino acid p-dist within genotype	Associated disease(s)	Reference(s)
AAstV 1[b]	Turkey (*Meleagris meleagris*)	Turkey astrovirus 1 (TAstV-1)	GI.A		Y15936	–	Diarrhea, decreased growth	[27, 54]
AAstV 2[b]	Chicken (*Gallus gallus*)	Avian nephritis virus (ANV)	GI.B	ANV-1	AB033998	0.299	Growth depression and kidney lesions	[24, 58]
	Turkey (*Meleagris meleagris*)			ANV-2	AB046864			
AAstV 3[b]	Turkey (*Meleagris meleagris*)	Turkey astrovirus 2 (TAstV-2)	GII.A	TAstV-2/1987	AY769615	0.181	Poult enteritis mortality syndrome (PEMS)	[32, 66, 67]
	Guinea fowl (*Numida meleagris*)			TAstV-3/2001	AY769616			[7, 8]
	Duck (*Anas platyrhynchos domestica*)	Duck astrovirus 1 (DAstV-1)			FJ434664		Fatal hepatitis in ducklings	[8]
AAstV 4	Chicken (*Gallus gallus*)	Chicken astrovirus (CAstV)	–		JF414802	–	Runting and stunting syndrome (RSS); mild diarrhea and distension of the small intestine in young chicks	[2, 30]
AAstV 5	Pigeon (*Columbia livia*)	Pigeon avian nephritis virus (Pi-ANV)	–		HQ889774	0.047	Diarrhea in ~2-week-old pigeons	[77, 78]
	Chicken (*Gallus gallus*)				HM029238		Asymptomatic infections in chicks	[76]
AAstV 6	Wood pigeon (*Columba palumbus*)	Wood pigeon astrovirus (WPiAstV)	–		FR727149	–	–	[33]
AAstV 7	Feral pigeon (*Columbia livia*)	Feral pigeon astrovirus (FPiAstV)	–		FR727146 FR727147 FR727148	0.291	–	[33]

[a]Genotype species are in italics. Genotypes in italics and bold are included in the Taxonomic Proposal submitted to the ICTV by the *Astroviridae* Study Group in 2010 [5]. Genotypes in italics but regular type are proposed in this chapter, following the same criteria
[b]Astrovirus species considered in the current Ninth ICTV Report [6]

would be classified into 33 genotype species (Table 6.2). Mean amino acid genetic distances (p-dist) between and within genotypes range between 0.368–0.781, and 0.–0.318, respectively. Phylogenetic analysis of the complete capsid shows that all genotypes which contain more than one sequence are supported by a high bootstrap value (Fig. 6.1).

Human Astroviruses

Four out of these 33 described genotypes correspond to human viruses which co-circulate in the population: *MAstV 1* (eight serotypes of classic HAstV), *MAstV 6* (HAstV-MLB), *MAstV 8* (HMO-A or VA2), and *MAstV 9* (HMO-B, and HMO-C or VA1). While the association of classic HAstV (*MAstV 1*) to gastroenteritis is clear, the role of the other three highly divergent "nonclassical" astroviruses in human health still remains unclarified. Although most of the novel viruses were originally isolated in samples from patients with diarrhea [13, 14, 15, 16, 17, 31], they have also been detected in healthy controls and children with nonpoliovirus acute flaccid paralysis [31]. In a recent case–control study of HAstV-MLB1, "classical" human astroviruses were associated with diarrhea, while MLB1 was not [21], suggesting that their pathogenicity is expressed in only a fraction of infections. Interestingly, HAstV-MLB2 has been detected in the plasma of a febrile child with upper respiratory signs [22], while HAstV Puget Sound (HAstV-PS), which is 95% similar at the amino acid level in ORF2 to HAstV-VA1, has been isolated from the brain of a child suffering from encephalitis [51], indicating that extraintestinal manifestations of astroviruses are not restricted to avian viruses. Finally, human astroviruses closely related to genotype *MAstV 9* have been also isolated from stool samples from a few patients with onset of celiac disease [60].

Within genotypes, serotypes are distributed based on antigenic properties. While all eight serotypes of "classic" HAstVs can be propagated in cell culture, it is now clear that they correspond to true different antigenic types. However, whether MLB1 and MLB2, and HMO-B and HMO-C, represent different serotypes within a given genotype deserves further investigations. While the well-established eight serotypes share 65–85% amino acid similarities, sequence

Fig. 6.1 Phylogenetic relationships within the family *Astroviridae*. The predicted amino acid sequences of the entire capsid polyprotein were aligned using CLUSTAL Omega [59]. The phylogenetic tree was generated using the neighbor-joining algorithm [56] implemented in the MEGA4 program [65]. The percentage of replicate trees in which the associated taxa clustered together in the bootstrap test (100 replicates) is shown next to the branches. The tree is drawn to scale, with branch lengths in the same units as those of the evolutionary distances (p-dist) used to infer the phylogenetic tree. All positions containing alignment gaps and missing data were eliminated only in pairwise sequence comparisons (pairwise deletion option). *Names* in *italics* represent genotype species. Host(s) of origin is indicated in *brackets*. Genotypes in *italics* and *bold* are included in the Taxonomic Proposals submitted to the ICTV by the *Astroviridae* Study Group in 2010 [4, 5]. Genotypes in *italics* but regular type are proposed in this chapter, following the same criteria

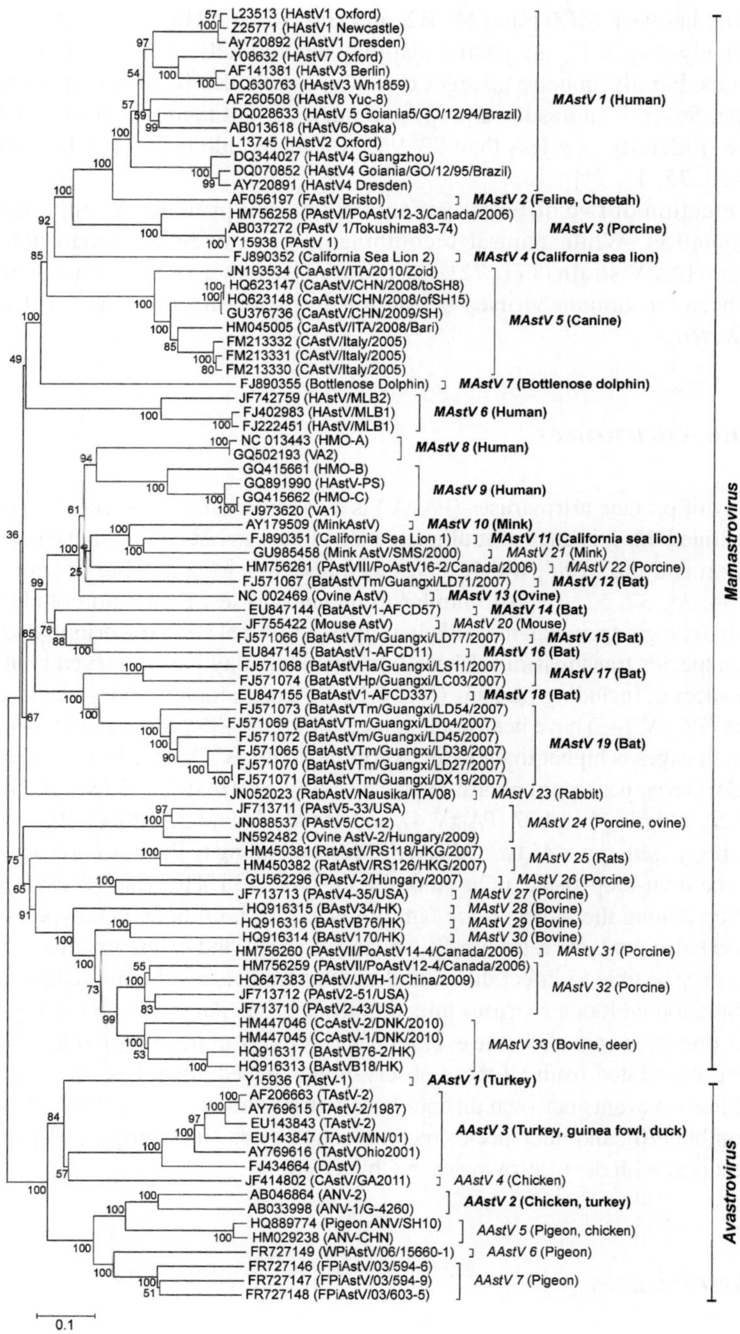
L23513 (HAstV1 Oxford)
Z25771 (HAstV1 Newcastle)
Ay720892 (HAstV1 Dresden)
Y08632 (HAstV7 Oxford)
AF141381 (HAstV3 Berlin)
DQ630763 (HAstV3 Wh1859)
AF260508 (HAstV8 Yuc-8)
DQ028633 (HAstV 5 Goiania5/GO/12/94/Brazil)
AB013618 (HAstV6/Osaka)
L13745 (HAstV2 Oxford)
DQ344027 (HAstV4 Guangzhou)
DQ070852 (HAstV4 Goiania/GO/12/95/Brazil)
AY720891 (HAstV4 Dresden)
MAstV 1 (Human)
AF056197 (FAstV Bristol)
MAstV 2 (Feline, Cheetah)
HM756258 (PAstVI/PoAstV12-3/Canada/2006)
AB037272 (PAstV 1/Tokushima83-74)
Y15938 (PAstV 1)
MAstV 3 (Porcine)
FJ890352 (California Sea Lion 2)
MAstV 4 (California sea lion)
JN193534 (CaAstV/ITA/2010/Zoid)
HQ623147 (CaAstV/CHN/2008/toSH8)
HQ623148 (CaAstV/CHN/2008/ofSH15)
GU376736 (CaAstV/CHN/2009/SH)
HM045005 (CaAstV/ITA/2008/Bari)
FM213332 (CAstV/Italy/2005)
FM213331 (CAstV/Italy/2005)
FM213330 (CAstV/Italy/2005)
MAstV 5 (Canine)
FJ890355 (Bottlenose Dolphin)
MAstV 7 (Bottlenose dolphin)
JF742759 (HAstV/MLB2)
FJ402983 (HAstV/MLB1)
FJ222451 (HAstV/MLB1)
MAstV 6 (Human)
NC 013443 (HMO-A)
GQ502193 (VA2)
MAstV 8 (Human)
GQ415661 (HMO-B)
GQ891990 (HAstV-PS)
GQ415662 (HMO-C)
FJ973620 (VA1)
MAstV 9 (Human)
AY179509 (MinkAstV)
MAstV 10 (Mink)
FJ890351 (California Sea Lion 1)
MAstV 11 (California sea lion)
GU985458 (Mink AstV/SMS/2000)
MAstV 21 (Mink)
HM756261 (PAstVIII/PoAstV16-2/Canada/2006)
MAstV 22 (Porcine)
FJ571067 (BatAstVTm/Guangxi/LD71/2007)
MAstV 12 (Bat)
NC 002469 (Ovine AstV)
MAstV 13 (Ovine)
EU847144 (BatAstV1-AFCD57)
MAstV 14 (Bat)
JF755422 (Mouse AstV)
MAstV 20 (Mouse)
FJ571066 (BatAstVTm/Guangxi/LD77/2007)
MAstV 15 (Bat)
EU847145 (BatAstV1-AFCD11)
MAstV 16 (Bat)
FJ571068 (BatAstVHa/Guangxi/LS11/2007)
FJ571074 (BatAstVHp/Guangxi/LC03/2007)
MAstV 17 (Bat)
EU847155 (BatAstV1-AFCD337)
MAstV 18 (Bat)
FJ571073 (BatAstVTm/Guangxi/LD54/2007)
FJ571069 (BatAstVTm/Guangxi/LD04/2007)
FJ571072 (BatAstVm/Guangxi/LD45/2007)
FJ571065 (BatAstVTm/Guangxi/LD38/2007)
FJ571070 (BatAstVTm/Guangxi/LD27/2007)
FJ571071 (BatAstVTm/Guangxi/DX19/2007)
MAstV 19 (Bat)
JN052023 (RabAstV/Nausika/ITA/08)
MAstV 23 (Rabbit)
JF713711 (PAstV5-33/USA)
JN088537 (PAstV5/CC12)
JN592482 (Ovine AstV-2/Hungary/2009)
MAstV 24 (Porcine, ovine)
HM450381(RatAstV/RS118/HKG/2007)
HM450382 (RatAstV/RS126/HKG/2007)
MAstV 25 (Rats)
GU562296 (PAstV-2/Hungary/2007)
MAstV 26 (Porcine)
JF713713 (PAstV4-35/USA)
MAstV 27 (Porcine)
HQ916315 (BAstV34/HK)
MAstV 28 (Bovine)
HQ916316 (BAstVB76/HK)
MAstV 29 (Bovine)
HQ916314 (BAstV170/HK)
MAstV 30 (Bovine)
HM756260 (PAstVII/PoAstV14-4/Canada/2006)
MAstV 31 (Porcine)
HM756259 (PAstVII/PoAstV12-4/Canada/2006)
HQ647383 (PAstV/JWH-1/China/2009)
JF713712 (PAstV2-51/USA)
JF713710 (PAstV2-43/USA)
MAstV 32 (Porcine)
HM447046 (CcAstV-2/DNK/2010)
HM447045 (CcAstV-1/DNK/2010)
HQ916317 (BAstVB76-2/HK)
HQ916313 (BAstVB18/HK)
MAstV 33 (Bovine, deer)
Y15936 (TAstV-1)
AAstV 1 (Turkey)
AF206663 (TAstV-2)
AY769615 (TAstV-2/1987)
EU143843 (TAstV-2)
EU143847 (TAstV/MN/01)
AY769616 (TAstVOhio2001)
FJ434664 (DAstV)
AAstV 3 (Turkey, guinea fowl, duck)
JF414802 (CAstV/GA2011)
AAstV 4 (Chicken)
AB046864 (ANV-2)
AB033998 (ANV-1/G-4260)
AAstV 2 (Chicken, turkey)
HQ889774 (Pigeon ANV/SH10)
HM029238 (ANV-CHN)
AAstV 5 (Pigeon, chicken)
FR727149 (WPiAstV/06/15660-1)
AAstV 6 (Pigeon)
FR727146 (FPiAstV/03/594-6)
FR727147 (FPiAstV/03/594-9)
FR727148 (FPiAstV/03/603-5)
AAstV 7 (Pigeon)
Mamastrovirus
Avastrovirus
0.1

similarity between MLB1 and MLB2, and HMO-B and HMO-C is 74% and 77%, respectively (Fig. 6.1), suggesting that they would likely show different antigenic properties. Finally, genetic lineages or subtypes can also be identified within each serotype. Several studies have proposed that a variant could be considered a new subtype if identity was less than 93–95% of nucleotide homology to a reference strain [20, 25, 42, 71].

Coinfection of two different variants in one cell provides an opportunity for recombination. While natural recombinants have been described among the classical HAstV strains [71, 72], the question whether recombination may also occur between human viruses belonging to different genotypes still remains unanswered.

Porcine Astroviruses

Diversity of porcine astroviruses (PAstV) is also remarkable, and recent reports have demonstrated the existence of multiple genetic lineages circulating in the swine community around the world, which serves as a reservoir for a number of zoonotic diseases [34, 35, 37, 52, 57, 70]. Similar to human viruses, PAstV do not cluster in a monophyletic group, which most likely suggests different ancestral origins and numerous interspecies transmissions. These transmissions may have involved many mammalian species, including humans (Fig. 6.1). To date, a total of five so-called genetic lineages (PAstV 1–5) have been documented for PAstV [35]. Since p-dist within some of these lineages is higher than 0.368, according to the ICTV criteria, they actually correspond to seven genotype species: *MAstV 3* (PAstV 1), *MAstV 22* (PAstV 3), *MAstV 24* (PAstV 5), *MAstV 26* and *27* (PAstV 4), and *MAstV 31* and *32* (PAstV 2) (Table 6.2). Interestingly, genotypes *MAstV 31* and *32*, corresponding to the most prevalent PAstV 2, lack the stem-loop-II-like (s2m) motif in the ORF2/3′UTR, which is usually highly conserved among the *Astroviridae* family [34, 37, 52]. Of note, genotype *MAstV 24* (PAstV 5) also contains astroviruses isolated in sheep, indicating that the same virus species may be able to infect different domestic animals and highlighting the risk of recombination with other viruses infecting these hosts. An active role of pigs as well as other domestic animals in the evolution and ecology of the *Astroviridae* family can easily be speculated from all these observations. Indeed, there has been evidence of recombination events between distinct PAstV strains and between PAstV and HAstV in Colombia [70], and interspecies recombination events involving PAstVs may have also occurred with deer astroviruses in China [34].

Bat Astroviruses

Bats constitute approximately 20% of living mammal species and are increasingly recognized as a reservoir of a number of human zoonotic diseases. Several reports from China [10, 75, 79] and one from Germany [12] have documented the identification of novel groups of astroviruses in apparently healthy bats with very high prevalences. Although insectivorous bats may be more likely to harbor astroviruses than frugivorous bats, BatAstVs have also been detected in one species of fruit bats [75]. Genetic diversity of bat astroviruses has been mostly analyzed in a partial region of the ORF1b gene. Within mamastroviruses, phylogenetic analysis of ORF1b gene results in seven monophyletic groups [79]. With the exception of group 1, BatAstVs fall in all remaining groups, and five groups are completely occupied by BatAstVs [79]. Although the complete capsid has not been sequenced for all BatAstV isolates, following the ICTV genetic criteria, available BatAstV sequences are classified in seven genotype species (*MAstV 12*, *MAstV 14*, *MAstV 15*, *MAstV 16*, *MAstV 17*, *MAstV 18*, and *MAstV 19*) [4]. Similar to what is observed for other mamastrovirus genotypes, some of them correspond to astroviruses infecting a single species of bats, while others have been isolated from more than one species, and viruses belonging to different genotypes have been detected in the same species of bats.

Bovine Astroviruses

Although bovine astroviruses (BAstVs) were discovered in 1978 [73], complete genome sequences of selected BAstV strains have not been available until 2011 [69]. Phylogenetic analysis of these sequences indicates that they are related to several porcine and deer (*Capreolus capreolus*) astroviruses [69], and a total of four genotype species with genetic distances greater than 0.368 can be identified: *MAstV 28*, *MAstV 29*, *MAstV 30*, and *MAstV 33* (Fig. 6.1). Of note, genotype *MAstV 33* would also include viruses isolated in European roe deer, suggesting that these BAstVs and *Capreolus capreolus* astroviruses (CcAstVs) may represent different strains of the same virus species which can infect different hosts, similar to the situation of feline and cheetah astroviruses [69]. Indeed, Tse and coworkers isolated for the first time different BAstV genotypes in the same host, and also reported that at least one recombination event had occurred between BAstVs and CcAstVs during the evolution of these viruses [69].

The application of cross-virus neutralization assays enabled many years ago the identification of at least two serotypes of BAstVs, but no sequence information was provided at that time [74]. Since genetic distances between current BAstV genotypes are considerably higher than those observed between viruses belonging to different human serotypes, it is likely that BAstV isolates reported to be different serotypes [74] were in fact viruses belonging to different genotypes. Based on genetic distance analysis of the two roe deer astroviruses isolated to date, it has been suggested that they also constitute two different antigenic variants [61].

Other Astrovirus Infecting Mammals

The use of consensus PCR techniques and the advent of viral metagenomic strategies have enabled the identification of other astroviruses infecting many other mammal species such as cats, dogs, sheep, minks, mice, rats, rabbits, cheetahs, and several species of marine mammals (California and Steller sea lions, and bottlenose dolphins) [11, 41, 50]. So far, feline, canine, mouse, rat, rabbit, California sea lion, and dolphin astroviruses variants isolated to date fall within single genotype species, while mink and sheep astroviruses contain representatives in separate clades or genotypes with likely independent origins (Fig. 6.1) (Table 6.2).

Regarding ovine astroviruses (OAstVs), two genetically divergent lineages have been identified. OAstV-1, corresponding to *MAstV 13,* represents the first report of the occurrence of astrovirus infections in animals other than humans [63], while OAstV-2 (included within genotype *MAstV 24*) is a recent novel isolate detected in sheep in Hungary [53]. Interestingly, *MAstV 24* genotype also includes porcine astroviruses, raising the question whether these virus species are able to infect both domestic sheep and pigs (Fig. 6.1) (Table 6.2).

Similarly to OAstVs, two genotypes of mink astroviruses have been identified. Besides indicating that astroviruses would have been introduced in minks more than once, these genotypes are also associated to different diseases. While genotype *MAstV 10* includes mink astroviruses that cause preweaning diarrhea syndrome [44], genotype *MAstV 21* likely represents the etiological agent of a neurological disorder, the "shaking mink syndrome" [3]. Whether differences in the capsid region of these viruses alter the cell tropism is not known, and consequently their disease spectrum is yet to be confirmed. Of note, *MAstV 21* shows a certain level of similarity to HAstV-PS, which belongs to genotype *MAstV 9* and was recently isolated from a child suffering from encephalitis [51]. Together with other astroviruses (human, California sea lion, porcine, bat, and ovine), these two isolates are grouped in a larger cluster with a 100 bootstrap value (Fig. 6.1).

Canine astroviruses (CaAstVs) are increasingly recognized as enteric pathogens in puppies, and isolates so far sequenced belong to genotype *MAstV 5* [39, 40, 68, 80]. Regarding feline astroviruses (FAstVs), genotype *MAstV 2* includes virus species infecting cats, and although cheetah astroviruses have only been partially sequenced, they are generally considered as the same virus species [1]. On the contrary, astroviruses found in marine mammals (*MAstV 4*, *MAstV 7*, and *MAstV 11*) are very diverse and some of them are similar to terrestrial astroviruses, suggesting that the marine environment has played an important role in astrovirus ecology (Fig. 6.1) [55]. Indeed, a recombination event may have occurred between a human and a California sea lion astrovirus, suggesting that both lineages may have been at one point capable of infecting the same host [55]. The analysis of the fecal viral flora of wild rodents led to the identification of the first murine astrovirus genome (*MAstV 20*) [50], and novel astroviruses have also been isolated in urban brown rats in Hong Kong (*MAstV 25*) [9]. Finally, a novel astrovirus has also been identified in rabbits (*MAstV 23*); whether these viruses play a role in rabbit enteritis complex or rabbit enterocolitis diseases still needs to be clarified [41].

Table 6.4 Tentative genotype species within the *Mamastrovirus* genera (pending complete sequencing of the capsid region)

Name of tentative genotype species	Accession number(s)
California sea lion astrovirus 1 (CslAstV 1)	FJ890351
California sea lion astrovirus 3 (CslAstV 3)	FJ890353
Steller sea lion astrovirus 1 (SslAstV 1)	FJ890354

Finally, astroviruses for which only partial sequences are available, or no sequences at all are deposited in Genbank, are listed as "Related viruses candidate to be members of the genus but have not been approved as species" (Table 6.4). In the near future, as complete capsid sequences are obtained, novel isolates may be assigned to newly described genotypes following the established criteria.

Astrovirus Infecting Birds (Genus *Avastrovirus*)

In avian hosts, astroviruses have been related to enteritis and poult enteritis mortality syndrome (PEMS) in young turkeys, chickens, and guinea fowls, to kidney lesions and nephritis in chickens, to the runting and stunting syndrome (RSS) in chickens, and to a fatal hepatitis in ducks [30, 49]. Recently, the screening of several species of pigeons harboring the conserved stem-loop-2-like (s2m) mobile viral RNA motif has resulted in the discovery of three novel pigeon astroviruses [33]. Finally, a novel virus related to the avian nephritis virus (ANV) has also been isolated and characterized in pigeons with gastroenteritis [77, 78], as well as in healthy chickens [76].

Thus, according to the new criteria proposed by the *Astroviridae* Study Group [5], avastroviruses are classified into seven genotype species (Table 6.3). Mean amino acid genetic distances (p-dist) between and within genotypes are similar to those reported for mamastroviruses and range between 0.414–0.728, and 0.047–0.299, respectively. Again, phylogenetic analysis of the complete capsid shows that all genotypes which contain more than one sequence are supported by a high bootstrap value (Fig. 6.1).

Turkey astroviruses (TAstVs) are widely spread. Although initially considered as two serotypes of the same virus, TAstV-1 and TAstV-2 are actually genetically unrelated types that resulted from separate introductions into turkeys. In general, TAstV-2 are more diverse and more prevalent than TAstV-1. In astrovirus taxonomy, these viruses correspond to genotypes *AAstV-3* and *AAstV-1*, respectively (Table 6.3). While it is clear that at least two serotypes of TAstV-2 exist [67], and possibly three based on percentage of sequence identity [64] (Fig. 6.1), no serological characterization of TAstV-1 isolates has been reported. Interestingly, recombination events have been reported between TAstVs circulating in turkey flocks [64]. Genotype *AAstV-3* also contains astroviruses infecting guinea fowl, which are closely related to TAstV-2 [7, 8]. Interspecies transmissions between turkeys and guinea fowls have been suggested as a possible mechanism by which these viruses may have evolved. Finally, astroviruses associated with fatal hepatitis in ducklings are also contained within *AAstV-3* genotype. From a phylogenetic point of view, it has been suggested that duck astroviruses may have originally been transmitted from turkeys [18].

Avian nephritis virus (ANV) is also widely distributed in chickens. According to the ICTV proposal, ANV correspond to genotype species *AAstV 2*. At least two serotypes of ANV have been reported [58], and ANV infections have also been described in turkeys, although these infections have never been associated with disease in these birds [48].

Besides ANV, chickens can also be infected by a second type of chicken astrovirus [2], which would correspond to genotype species *AAstV 4* (Table 6.3). Phylogenetic analysis shows that, despite infecting the same host as ANV, these chicken astroviruses (CAstVs) are more closely related to TAstV-1 and TAstV-2 than to ANV (Fig. 6.1). Of note, determination of the full-length sequence of the first CAstV belonging to *AAstV 4* revealed that the replication mechanism of this virus may be different from that of all other studied astroviruses, with ORF1b having its own start codon [30].

Finally, although pigeons were considered unsusceptible to some viral infections that induced significant diseases in poultry, a recent survey performed in China among pigeon flocks with diarrhea allowed the first identification of pigeon ANV [77]. Interestingly, this pigeon virus is closely related to a strain of ANV isolated in healthy chicken flocks in the same country [76]. Following the ICTV criteria for astrovirus classification, and also suggested by the authors of the study [77], these isolates constitute the same virus species. Thus, *AAstV 5* genotype represents another example of a single virus species infecting at least two different hosts (Table 6.3). Of note, pigeon ANV is not the only astrovirus infecting pigeons. Genotypes *AAstV 6* and *AAstV 7* also include pigeon astroviruses (PiAstVs) isolated recently from feces of wood and feral pigeons, respectively (Table 6.3) [32, 33]. Together with *AAstV 5* viruses, these genotypes strongly cluster together with classical ANV (Fig. 6.1).

Conclusions

In conclusion, all these new astrovirus descriptions indicate that astroviruses have a broader spectrum of host species as well as higher genetic and antigenic diversity than previously anticipated. While they had been historically regarded as species-specific viruses, it is now evident that they are a remarkably genetically diverse family with a great potential for colonization of different environments. Members of multiple astrovirus species can infect the same host, while other variants can infect more than one animal species, denoting weaker host restriction factors. Together with the high tendency towards the incorporation of mutations and the occurrence of recombination events, these observations set the stage for interspecies transmission and virus adaptations to new hosts. Although astrovirus infections are subclinical in many hosts, their notable genetic diversity, their tendency to recombine, and their ability to cross species barriers make them important candidates for emerging infections of zoonotic origin. Viral surveys of human and animals, as well as regular updates of their taxonomy, are needed to facilitate detection of these emerging zoonotic viruses, as well as to evaluate their pathogenic potential.

References

1. Atkins A, Wellehan Jr JF, Childress AL, Archer LL, Fraser WA, Citino SB. Characterization of an outbreak of astroviral diarrhea in a group of cheetahs (*Acinonyx jubatus*). Vet Microbiol. 2009;14:160–5.
2. Baxendale W, Mebatsion T. The isolation and characterisation of astroviruses from chickens. Avian Pathol. 2004;33:364–70.
3. Blomström AL, Widén F, Hammer AS, Belák S, Berg M. Detection of a novel astrovirus in brain tissue of mink suffering from shaking mink syndrome by use of viral metagenomics. J Clin Microbiol. 2010;48:4392–6.
4. Bosch A, Guix S, Krishna NK, Méndez E, Monroe SS, Pantin-Jackwood M, Schultz-Cherry S. Nineteen new species in the genus *Mamastrovirus* in the *Astroviridae* family. 2010a. ICTV 2010.018a-cV.A.v4.
5. Bosch A, Guix S, Krishna NK, Méndez E, Monroe SS, Pantin-Jackwood M, Schultz-Cherry S. Three new species in the genus *Avastrovirus* in the *Astroviridae* family. 2010b. ICTV 2010.017abV.A.v2.
6. Bosch A, Méndez E, Krishna NK, Pantin-Jackwood M, Schultz-Cherry S, Monroe SS, et al. *Astroviridae*. In: King AMQ, Adams MJ, Carstens EB, Lefkowitz EJ, editors. Virus taxonomy. 9th ed. Oxford: Elsevier; 2011.
7. Cattoli G, De Battisti C, Toffan A, Salviato A, Lavazza A, Cerioli M, et al. Co-circulation of distinct genetic lineages of astroviruses in turkeys and guinea fowl. Arch Virol. 2007;152:595–602.
8. Cattoli G, Toffan A, De Battisti C, Salviato A, Terregino C, Capua I. Astroviruses found in the intestinal contents of guinea fowl suffering from enteritis. Vet Rec. 2005;156:220.
9. Chu DKW, Chin AWH, Smith GJ, Chan KH, Guan Y, Peiris JSM, et al. Detection of novel astroviruses in urban brown rats and previously known astroviruses in humans. J Gen Virol. 2010;91:2457–62.
10. Chu DKW, Poon LLM, Guan Y, Peiris JSM. Novel astroviruses in insectivorous bats. J Virol. 2008;82:9107–14.
11. De Benedictis P, Schultz-Cherry S, Burnham A, Cattoli G. Astrovirus infections in humans and animals - molecular biology, genetic diversity, and interspecies transmissions. Infect Genet Evol. 2011;11:1529–44.
12. Drexler JF, Corman VM, Wegner T, Tateno AF, Zerbinati RM, Gloza-Rausch F, et al. Amplification of emerging viruses in a bat colony. Emerg Infect Dis. 2011;17:449–56.
13. Finkbeiner SR, Allred AF, Tarr PI, Klein EJ, Kirkwood CD, Wang D. Metagenomic analysis of human diarrhea: viral detection and discovery. PLoS Pathog. 2008;4:e1000011.
14. Finkbeiner SR, Kirkwood CD, Wang D. Complete genome sequence of a highly divergent astrovirus isolated from a child with acute diarrhea. Virol J. 2008;5:117.
15. Finkbeiner SR, Le BM, Holtz LR, Storch GA, Wang D. Detection of newly described astrovirus MLB1 in stool samples from children. Emerg Infect Dis. 2009;15:441–4.

16. Finkbeiner SR, Li Y, Ruone S, Conrardy C, Gregoricus N, Toney D, et al. Identification of a novel astrovirus (astrovirus VA1) associated with an outbreak of acute gastroenteritis. J Virol. 2009;83:10836–9.
17. Finkbeiner SR, Holtz LR, Jiang Y, Rajendran P, Franz CJ, Zhao G, et al. Human stool contains a previously unrecognized diversity of novel astroviruses. Virol J. 2009;6:161.
18. Fu Y, Pan M, Wang X, Xu Y, Xie X, Knowles NJ, et al. Complete sequence of a duck astrovirus associated with fatal hepatitis in ducklings. J Gen Virol. 2009;90:1104–8.
19. Fuentes, C, Bosch A, Pintó RM, Guix S. Identification of human astrovirus genome-linked protein (VPg) essential for viral infectivity. J Virol. 2012 (in press).
20. Guix S, Caballero S, Villena C, Bartolomé R, Latorre C, Rabella N, et al. Molecular epidemiology of astrovirus infection in Barcelona Spain. J Clin Microbiol. 2002;40:133–9.
21. Holtz LR, Bauer IK, Rajendran P, Kang G, Wang D. Astrovirus MLB1 is Not associated with diarrhea in a cohort of Indian children. PLoS One. 2011;6:e28647.
22. Holtz LR, Wylie KM, Sodergren E, Jiang Y, Franz CJ, Weinstock GM, et al. Astrovirus MLB2 viremia in febrile child. Emerg Infect Dis. 2011;17:2050–2.
23. Hoshino Y, Zimmer JF, Moise NS, Scott FW. Detection of astroviruses in feces of a cat with diarrhea. Brief report. Arch Virol. 1981;70:373–6.
24. Imada T, Yamaguchi S, Mase M, Tsukamoto K, Kubo M, Morooka A. Avian nephritis virus (ANV) as a new member of the family *Astroviridae* and construction of infectious ANV cDNA. J Virol. 2000;74:8487–93.
25. Jakab F, Walter JE, Berke T, Matson DO, Mitchell DK, Szucs G. Molecular characterization and sequence analysis of human astroviruses circulating in Hungary. FEMS Immunol Med Microbiol. 2003;39:97–102.
26. Jiang B, Monroe SS, Koonin EV, Stine SE, Glass RI. RNA sequence of astrovirus: distinctive genomic organization and a putative retrovirus-like ribosomal frameshifting signal that directs the viral replicase synthesis. Proc Natl Acad Sci USA. 1993;90:10539–43.
27. Jonassen CM, Jonassen TO, Grinde B. A common RNA motif in the 3′ end of the genomes of astroviruses, avian infectious bronchitis virus and an equine rhinovirus. J Gen Virol. 1998;79:715–8.
28. Jonassen CM, Jonassen TO, Saif YM, Snodgrass DR, Ushijima H, Shimizu M, et al. Comparison of capsid sequences from human and animal astroviruses. J Gen Virol. 2001;82:1061–7.
29. Jonassen CM, Jonassen TTO, Sveen TM, Grinde B. Complete genomic sequences of astroviruses from sheep and turkey: comparison with related viruses. Virus Res. 2003;91:195–201.
30. Kang KI, Icard AH, Linnemann E, Sellers HS, Mundt E. Determination of the full-length sequence of a chicken astrovirus suggests a different replication mechanism. Virus Genes. 2011;44(1):45–50.
31. Kapoor A, Li L, Victoria J, Oderinde B, Mason C, Pandey P, et al. Multiple novel astrovirus species in human stool. J Gen Virol. 2009;90:2965–72.
32. Koci MD, Seal BS, Schultz-Cherry S. Molecular characterization of an avian astrovirus. J Virol. 2000;74:6173–7.
33. Kofstad T, Jonassen CM. Screening of feral and wood pigeons for viruses harbouring a conserved mobile viral element: characterization of novel astroviruses and picornaviruses. PLoS One. 2011;6:e25964.
34. Lan D, Ji W, Shan T, Cui L, Yang Z, Yuan C, et al. Molecular characterization of a porcine astrovirus strain in China. Arch Virol. 2011;156:1869–75.
35. Laurin MA, Dastor M, L'homme Y. Detection and genetic characterization of a novel pig astrovirus: relationship to other astroviruses. Arch Virol. 2011;156:2095–9.
36. Lewis TL, Greenberg HB, Herrmann JE, Smith LS, Matsui SM. Analysis of astrovirus serotype 1 RNA, identification of the viral RNA-dependent RNA polymerase motif, and expression of a viral structural protein. J Virol. 1994;68:77–83.
37. Luo Z, Roi S, Dastor M, Gallice E, Laurin MA, L'homme Y. Multiple novel and prevalent astroviruses in pigs. Vet Microbiol. 2011;149:316–23.

38. Madeley CR, Cosgrove BP. Letter: 28 nm particles in faeces in infantile gastroenteritis. Lancet. 1975;2(7932):451–2.
39. Martella V, Moschidou P, Catella C, Larocca V, Pinto P, Losurdo M, et al. Enteric disease in dogs naturally infected by a novel canine astrovirus. J Clin Microbiol. 2012;50(3):1066–9.
40. Martella V, Moschidou P, Lorusso E, Mari V, Camero M, Bellacicco A, et al. Detection and characterization of canine astroviruses. J Gen Virol. 2011;92:1880–7.
41. Martella V, Moschidou P, Pinto P, Catella C, Desario C, Larocca V, et al. Astroviruses in rabbits. Emerg Infect Dis. 2011;17:2287–93.
42. Medina SM, Gutierrez MF, Liprandi F, Ludert JE. Identification and type distribution of astroviruses among children with gastroenteritis in Colombia and Venezuela. J Clin Microbiol. 2000;38:3481–3.
43. Méndez E, Arias DF. *Astroviridae*. In: Knipe DM, Howley PM, Griffin DE, Lamb RA, Martin MA, Roizman B, Straus SE, editors. Fields virology. Philadelphia: Lippincott Williams & Wilkins; 2007.
44. Mittelholzer C, Hedlund KO, Englund L, Dietz HH, Svensson L. Molecular characterization of a novel astrovirus associated with disease in mink. J Gen Virol. 2003;84:3087–94.
45. Monroe SS, Carter MJ, Herrmann JE, Kurtz JB, Matsui SM. *Astroviridae*. In: Murphy FA, Fauquet CM, Bishop DHL, Ghabrail SA, Jarvis AW, Martielli GP, Mayo MA, Summers MD, editors. Virus taxonomy: classification and nomenclature of viruses, sixth report of the international committee on taxonomy of viruses. Wien, Austria: Springer; 1995.
46. Monroe SS, Jiang B, Stine SE, Koopmans M, Glass RI. Subgenomic RNA sequence of human astrovirus supports classification of *Astroviridae* as a new family of RNA viruses. J Virol. 1993;67:3611–4.
47. Monroe SS, Stine SE, Gorelkin L, Herrmann JE, Blacklow NR, Glass RI. Temporal synthesis of proteins and RNAs during human astrovirus infection of cultured cells. J Virol. 1991;65:641–8.
48. Pantin-Jackwood MJ, Spackman E, Woolcock PR. Molecular characterization and typing of chicken and turkey astroviruses circulating in the United States: implications for diagnostics. Avian Dis. 2;50:397–404.
49. Pantin-Jackwood MJ, Strother KO, Mundt E, Zsak L, Day JM, Spackman E. Molecular characterization of avian astroviruses. Arch Virol. 2011;156(2):235–44.
50. Phan TG, Kapusinszky B, Wang C, Rose RK, Lipton HL, Delwart EL. The fecal viral flora of wild rodents. PLoS Pathog. 2011;7(9):e1002218.
51. Quan PL, Wagner TA, Briese T, Torgerson TR, Hornig M, Tashmukhamedova A, et al. Astrovirus encephalitis in boy with X-linked agammaglobulinemia. Emerg Infect Dis. 2010;16:918–25.
52. Reuter G, Pankovics P, Boros A. Identification of a novel astrovirus in a domestic pig in Hungary. Arch Virol. 2011;156:125–8.
53. Reuter G, Pankovics P, Delwart E, Boros A. Identification of a novel astrovirus in domestic sheep in Hungary. Arch Virol. 2012;157(2):323–7.
54. Reynolds DL, Saif YM. Astrovirus: a cause of an enteric disease in turkey poults. Avian Dis. 1986;30:728–35.
55. Rivera R, Nollens HH, Venn-Watson S, Gulland FM, Wellehan Jr JF. Characterization of phylogenetically diverse astroviruses of marine mammals. J Gen Virol. 2010;91:166–73.
56. Saitou N, Nei M. The neighbor-joining method: a new method for reconstructing phylogenetic trees. Mol Biol Evol. 1987;4:406–25.
57. Shan T, Li L, Simmonds P, Wang C, Moeser A, Delwart E. The fecal virome of pigs on a high-density farm. J Virol. 2011;85:11697–708.
58. Shirai J, Nakamura K, Shinohara K, Kawamura H. Pathogenicity and antigenicity of avian nephritis isolates. Avian Dis. 1991;35:49–54.
59. Sievers F, Wilm A, Dineen D, Gibson TJ, Karplus K, Li W, et al. Fast, scalable generation of high-quality protein multiple sequence alignments using clustal omega. Mol Syst Biol. 2011;7:539.

60. Smits SL, van Leeuwen M, van der Eijk AA, Fraaij PL, Escher JC, Simon JH, et al. Human astrovirus infection in a patient with new-onset celiac disease. J Clin Microbiol. 2010;48:3416–8.
61. Smits SL, Leeuven M, Kuiken T, Hammer AS, Simon JH, Osterhaus ADME. Identification and characterization of deer astroviruses. J Gen Virol. 2010;91:2719–22.
62. Snodgrass DR, Angus KW, Gray EW, Menzies JD, Paul G. Pathogenesis of diarrhoea caused by astrovirus infections in lambs. Arch Virol. 1979;60:217–26.
63. Snodgrass DR, Gray EW. Detection and transmission of 30 nm virus particles (astroviruses) in faeces of lambs with diarrhoea. Arch Virol. 1977;55:287–91.
64. Strain E, Kelley LA, Schultz-Cherry S, Muse SV, Koci MD. Genomic analysis of closely related astroviruses. J Virol. 2008;82:5099–103.
65. Tamura K, Dudley J, Nei M, Kumar S. MEGA4: molecular evolutionary genetics analysis (MEGA) software version 4.0. Mol Biol Evol. 2007;24:1596–9.
66. Tang Y, Murgia AM, Saif YM. Molecular characterization of the capsid gene of two serotypes of turkey astroviruses. Avian Dis. 2005;49:514–9.
67. Tang Y, Saif YM. Antigenicity of two turkey astrovirus isolates. Avian Dis. 2004;48: 896–901.
68. Toffan A, Jonassen CM, De Battisti C, Schiavon E, Kofstad T, Capua I, et al. Genetic characterization of a new astrovirus detected in dogs suffering from diarrhoea. Vet Microbiol. 2009;139:147–52.
69. Tse H, Chan WM, Tsoi HW, Fan RY, Lau CC, Lau SK, et al. Rediscovery and genomic characterization of bovine astroviruses. J Gen Virol. 2011;92:1888–98.
70. Ulloa JC, Gutiérrez MF. Genomic analysis of two ORF2 segments of new porcine astrovirus isolates and their close relationship with human astroviruses. Can J Microbiol. 2010;56: 569–77.
71. Walter JE, Briggs J, Guerrero ML, Matson DO, Pickering LK, Ruiz-Palacios G, et al. Molecular characterization of a novel recombinant strain of human astrovirus associated with gastroenteritis in children. Arch Virol. 2001;146:2357–67.
72. Wolfaardt M, Kiulia NM, Mwenda JM, Taylor MB. Evidence of a recombinant wild-type human astrovirus strain from a Kenyan child with gastroenteritis. J Clin Microbiol. 2011;49:728–31.
73. Woode GN, Bridger JC. Isolation of small viruses resembling astroviruses and caliciviruses from acute enteritis of calves. J Med Microbiol. 1978;11:441–52.
74. Woode GN, Gourley NE, Pohlenz JF, Liebler EM, Mathews SL, Hutchinson MP. Serotypes of bovine astrovirus. J Clin Microbiol. 1985;22:668–70.
75. Xiao J, Li J, Hu G, Chen Z, Wu Y, Chen Y, et al. Isolation and phylogenetic characterization of bat astroviruses in southern china. Arch Virol. 2011;156:1415–23.
76. Zhao W, Hua XG, Yuan L, Cui L, Shan TL, Dai XQ, et al. Sequence analyses of the representative Chinese-prevalent strain of avian nephritis virus in healthy chicken flocks. Avian Dis. 2011;55:65–9.
77. Zhao W, Zhu AL, Yu Y, Yuan CL, Zhu CX, Yang ZB, et al. Complete sequence and genetic characterization of pigeon avian nephritis virus, a member of the family *Astroviridae*. Arch Virol. 2011;156:1559–65.
78. Zhao W, Zhu AL, Yuan CL, Yu Y, Zhu CX, Lan DL, et al. Detection of astrovirus infection in pigeons (*Columbia livia*) during an outbreak of diarrhoea. Avian Pathol. 2011;40:361–5.
79. Zhu HC, Chu DK, Liu W, Dong BQ, Zhang SY, Zhang JX, et al. Detection of diverse astroviruses from bats in China. J Gen Virol. 2009;90:883–7.
80. Zhu AL, Zhao W, Yin H, Shan TL, Zhu CX, Yang X, et al. Isolation and characterization of canine astrovirus in China. Arch Virol. 2011;156:1671–5.

Chapter 7
New Human Astroviruses

Stacy R. Finkbeiner and Lori R. Holtz

Abstract The first human astrovirus was discovered in 1975 by Madeley and Cosgrove through visualization of viral particles in stool using electron microscopy. Over the course of the next ~20 years, an additional seven serotypes of human astroviruses were discovered. In the last decade, new technological advances in microarrays and sequencing strategies enabled more sophisticated methods for detecting viruses in clinical and environmental specimens. These methods led to the discovery of many novel viruses from a variety of virus families. They also brought about a dramatic realization that more astroviruses can be found in humans than previously recognized. In fact, the number of astroviruses associated with humans has nearly doubled within the last few years. Furthermore, the discovery of novel astroviruses in human specimens revealed that there is more diversity amongst them than was assumed based on the high level of similarity between human astroviruses 1 and 8. This chapter will describe the discovery and early characterization of the five novel astroviruses initially identified in human stool samples in 2008 and 2009.

S.R. Finkbeiner, Ph.D. (✉)
Department of Molecular Virology and Microbiology, Baylor College of Medicine,
One Baylor Plaza, Houston, TX 77030, USA
e-mail: stacy.finkbeiner@gmail.com

L.R. Holtz, M.D.
Department of Pediatrics, Washington University School of Medicine/St. Louis Children's Hospital, Children's Place, St. Louis, MO 63110, USA
e-mail: holtz_l@kids.wustl.edu

S. Schultz-Cherry (ed.), *Astrovirus Research: Essential Ideas, Everyday Impacts, Future Directions*, DOI 10.1007/978-1-4614-4735-1_7,

Introduction

The first human astrovirus was discovered in 1975 by Madeley and Cosgrove [34]. Over the course of the next ~20 years, an additional seven serotypes of human astroviruses were discovered [5, 21, 29, 31, 32]. However, new technological advances in virus discovery and detection methods have brought about a dramatic realization that more astroviruses can be found in humans than previously recognized and that there is more diversity amongst them than was assumed based on comparisons of human astroviruses 1–8. This chapter will describe the discovery and early characterization of the five novel astroviruses initially identified in human stool samples in 2008 and 2009.

Novel Virus Discovery Methods

Based on the tenets of Koch's postulates [17], the ability to grow a virus was long held as an essential step in its identification. However, enteric viruses have been notoriously difficult to grow in culture and in fact this has been the biggest ongoing challenge for the last 40 years in the human norovirus field [11]. The successful cultivation of some enteric viruses has depended on the inclusion of additives to the cultures (trypsin, pancreatin, bile, etc.) and/or serial blind passages in culture [3, 6, 49]. Even so, the ability to cultivate a virus does not inherently lead to its identification and characterization. In fact, visualization of viruses by electron microscopy (EM) was long used for the initial characterization of viruses that had been grown in culture or those present in stool samples from patients with diarrhea. In the 1970s, these methods were used to identify viral causes of diarrhea. The four families of viruses most commonly associated with diarrhea (*Rotaviridae, Caliciviridae, Astroviridae,* and *Adenovirus*) were all identified using EM [10, 18, 26, 34]. The identification of human astrovirus 1 was based on the observation of 28 nm virus particles present in stool from a child with gastroenteritis [34]. While these methods led to major advances in our understanding of the causes of diarrhea, they have limited our ability to detect a large number of viruses due to the fact that many viruses do not grow well, or at all, in cell culture and due to the relatively low sensitivity of EM studies. It is often difficult to classify a virus based on EM observations because many viral families exhibit similar morphologies under the microscope and also only 10% of astrovirus particles exhibit the classical star-like morphology [37]. Furthermore, the use of serological assays to determine the identity of an unknown virus is limited by the need for the unknown virus to cross-react with specific reagents for an already recognized virus. The use of serological assays is also hampered by the requirement for *a priori* knowledge of the possible identity of the unknown virus since there are currently no pan-viral serological assays available.

PCR-Based Strategies for Novel Virus Detection

The age of molecular biology has had an impact on virology just as it has on many other fields. In particular, the invention of the polymerase chain reaction (PCR) in the 1980s made it easy to detect viral genomes. If the appropriate sequences are available, it is quite easy to design primers to detect a given virus of interest. Analogous to the use of primers that target 16S ribosomal genes of bacteria for identification and classification of bacterial species, primers can often be designed to highly conserved regions of the genomes of viruses from a particular family or lower taxa. Such PCR assays can be used to identify novel viruses which happen to be closely related to existing viruses. Various computational strategies have been described for the specific design of consensus PCR primers to use for novel virus identification [24, 45]. The breadth of PCR assays is limited to interrogating a single taxon, which can be a limitation when there is no information to focus the efforts towards a particular group of viruses. However, there are numerous examples of viruses discovered in this fashion, with two notable instances being the identification of sin nombre virus [40] and human coronavirus HKU1 [53]. As will be discussed later in this chapter, PCR assays that were designed after the discovery of novel astrovirus MLB1 were used to identify additional novel astroviruses.

DNA Sequencing Methods for Novel Virus Detection

Dideoxy sequencing or Sanger sequencing, first described in 1977 [46], was the dominant DNA sequencing technology used for nearly 30 years. The typical sequence read length for Sanger sequencing can reach ~750–800 bp. Sanger sequencing is carried out individually for each sequence, so the number of sequences that can be generated for a single sample is simply a function of cost and time. This technology was the cornerstone of all significant sequencing projects up until ~2005, including the sequencing of the human genome. Starting in 2005, multiple new sequencing technology platforms ("NextGen") emerged that have now far surpassed conventional Sanger sequencing in terms of increased total sequence production capacity and decreased cost. With NextGen platforms, sequences can be modified to encode specific strings of bases that serve as barcodes for individual samples, such that many samples can be combined into a single run on the NextGen machines. The NextGen platform that has been most widely used for virus detection and sequencing is 454 pyrosequencing [44]. At the time of the studies described in this chapter, one 454 sequencing run produced ~400,000 sequence reads of an average read length of ~250–400 bp [35]. Both Sanger sequencing and 454 pyrosequencing have been used in the discovery and sequencing of novel astroviruses.

Novel Astroviruses Found in Human Biological Samples

Astrovirus MLB1

Human rotaviruses and noroviruses are the leading causes of viral diarrhea [9, 42, 52] while human astroviruses have been associated with up to ~10% of sporadic cases of viral diarrhea in children [19, 51] and with 0.5–15% of outbreaks [2, 33, 48]. However, the cause of approximately 12–41% of the cases of sporadic diarrhea and gastroenteritis outbreaks remains undetermined despite extensive testing [25, 33, 48]. One hypothesis for the cause of such a diagnostic gap is that there are unidentified pathogens contributing to the burden of disease. This was the guiding principle of the initial study that led to the identification of Astrovirus MLB1 (AstV-MLB1) [12]. Stool samples from children with cases of acute, sporadic diarrhea were subjected to "micro-mass sequencing" meaning that only limited sequencing was done (384 sequences) per sample in contrast to the typically tens of thousands of sequences generated per sample using NextGen sequencing platforms that were used in studies described later in this chapter. One of the most interesting findings of the study was the detection of sequences, within one sample collected in 1999 from a 35-month-old child, that were most closely related to human astroviruses. These sequences were highly divergent from canonical human astroviruses 1–8, suggesting that a novel virus might be present. There were seven unique sequence reads derived from two loci that displayed 42–61% amino acid identity to human astroviruses. The sequence reads overlapped such that they could be assembled into two contiguous stretches of sequence referred to as contigs. RT-PCR was used to confirm the accuracy of the assembled contigs and subsequent sequencing of the amplicons confirmed the presence of the contigs in the original RNA extract. One contig was ~800 bp long and mapped to ORF1a, the serine protease, of the astrovirus genome and the other one was ~500 bp long and mapped to ORF1b, the RNA polymerase. Phylogenetic analysis of the two contigs yielded trees that demonstrated that these sequences in fact represented a novel astrovirus (Fig. 7.1).

This novel astrovirus was named Astrovirus MLB1 (AstV-MLB1) because it was identified in a stool sample from a child in Melbourne, Australia. In the follow-up study, the complete genome of AstV-MLB1 was sequenced to an average of >3X coverage [GenBank: FJ222451] and the virus was further characterized [14]. The AstV-MLB1 genome is organized like that of other astroviruses in that it is predicted to encode three open reading frames (ORF1a, ORF1b, and ORF2) and contain both 5′ and 3′ nontranslated regions (NTR), as well as a poly-A tail. AstV-MLB1 has a shorter genome than other astroviruses with only 6,171 bp (excluding the poly-A tail) compared to the range for the other astroviruses of ~6,400–7,300 bp [37] (Table 7.1). This is in large due to two deletions, totaling 57 amino acids, located within a highly conserved 144 amino acid motif near the carboxyl terminus of human astroviruses 1–8. The motif has been mapped as an immunoreactive epitope in human astroviruses [36] and is located in the nonstructural protein p38 [28], a protein reported to be involved in replication [20] and

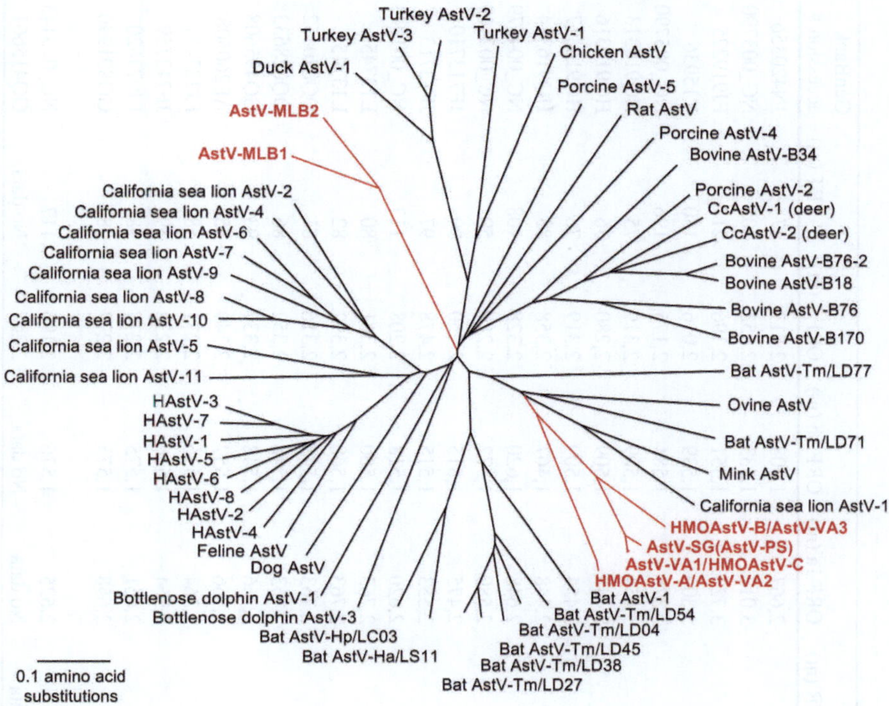

Fig. 7.1 Phylogenetic relationship of animal and human astroviruses based on capsid sequences. The neighbor-joining tree was generated using ClustalX2 [30] with 1,000 bootstrap replicates. *HAstV* human astrovirus. Novel astroviruses recently discovered in humans are shown in *red*

particle release [38]. The effect this deletion has on the replication of AstV-MLB1 has yet to be determined. Another difference between AstV-MLB1 and the canonical human astroviruses arises in the 3 C-like protease motif in the putative substrate binding residues. The canonical human astroviruses share the conserved residues RTQ whereas AstV-MLB1 is identical to ovine astrovirus with the residues ATR. Again, the meaning of this difference is yet to be explored.

AstV-MLB1 further differs from the canonical human astroviruses in some of its noncoding features. The 5′ nontranslated region (NTR) is only 14 nucleotides long which is more comparable to the 5′ NTRs of avian astroviruses (~10–20 nt) than to the 5′ NTRs of human astroviruses 1–8 (80–85 nt long, 37) (Table 7.1). Furthermore, human astroviruses 1–8 contain a 120 nt region at the junction between ORF1b and ORF2 that is highly conserved with a 52 nt core region that is 99–100% identical among the human astrovirus serotypes [50]. It is hypothesized that this sequence serves as a regulatory element of the sub-genomic RNA that encodes the capsid. AstV-MLB1 is only 61.5% identical to other human astroviruses in the 52 nt core region. This is similar to the level of identity (44–59.6%) shared between the known animal astroviruses and the human astroviruses. Lastly, there is a conserved RNA secondary

Table 7.1 Genome Information of fully sequenced astroviruses

Virus	Alternate names	Genome (nt)	5′ NTR (nt)	ORF 1a (nt)	ORF 1b (nt)	ORF 2 (nt)	3′ NRT (nt)	GenBank accession #
California sea lion AstV-9	None	6,627	32	2,667	1,608	2,517	47	JN420356
Chicken AstV-1	Avian nephritis virus	6,927	15	3,017	1,533	2,052	305	NC_003790
Duck AstV-1	Duck hepatitis virus 2	7,722	22	3,723	1,551	2,196	217	FJ919225
Turkey AstV-1	None	7,003	11	3,300	1,539	2,016	130	Y15936
Turkey AstV-2	None	7,325	21	3,378	1,584	2,175	196	NC_005790
Bovine AstV-B18	None	6,300	47	2,454	1,509	2,319	75	HQ916313
Bovine AstV-B76	None	6,253	44	2,448	1,509	2,280	76	HQ916316
Bovine AstV-B76-2	None	6,301	47	2,454	1,509	2,319	76	HQ916317
Bovine AstV-B170	None	6,317	23	2,448	1,467	2,358	83	HQ916314
Mink AstV	None	6,610	26	2,648	1,620	2,328	108	NC_004579
Ovine AstV	None	6,440	45	2,580	1,572	2,289	59	NC_002469
Porcine AstV-2	None	6,318	39	2,475	1,515	2,250	99	JF713710
Porcine AstV-4	None	6,639	74	2,553	1,515	2,478	97	JF13713
Porcine AstV-5	None	6,478	79	2,620	1,518	2,208	117	NC_001943
Human AstV-1	None	6,813	85	2,763	1,560	2,361	80	L13745
Human AstV-2	None	6,828	82	2,763	1,560	2,392	82	L13745
Human AstV-4	None	6,723	84	2,763	1,548	2,316	81	DQ070852
Human AstV-5	None	6,762	83	2,763	1,548	2,352	86	DQ028633
Human AstV-6	None	6,757	82	2,766	1,548	2,337	93	GQ495608
Human AstV-8	None	6,759	83	2,766	1,557	2,349	85	AF260508
AstV-MLB1	None	6,171	14	2,364	1,536	2,271	58	FJ222451
AstV-MLB2	None	6,119	14	2,364	1,536	2,238	39	JF742759
AstV-VA1	HMOAstV-C, HastV-PS	6,586	38	2,661	1,575	2,277	98	FJ973620
HastV-PS (HastV-SG in GenBank)	AstV-VA1, HMOAstV-C	6,584	39	2,658	1,575	2,277	98	GQ891990
HMOAstV-A	AstV-VA2	6,534	42	2,625	1,576	2,196	117	NC_013443
HMOAstV-B	AstV-VA3	No data	No data	No data	No data	2,267	No data	GQ415661

Genome information was obtained and/or deduced for all astroviruses in which full-length sequences were available at GenBank as of February 2012 structure, referred to as the stem-loop II-like motif (s2m), that is in the nontranslated region (NTR) at the 3′ end of the genome in all previously described astroviruses except turkey astrovirus 2 [39]. It is thought to interact with viral and cellular proteins needed for RNA replication and to be quite important because (1) it is also found in some other RNA viruses and (2) compensatory mutations that restore base pairing are often seen when there are mutations in the conserved motif [39]. AstV-MLB1 not only lacks this conserved nucleotide motif but it also has the shortest 3′NTR reported to date for an astrovirus. It is not clear the importance of any of these differences in the coding or noncoding regions of the AstV-MLB1 genome compared to other astroviruses, but these may be interesting avenues of research in the future.

Astrovirus MLB2

The primer pair that has predominantly been used to identify astroviruses in stool samples, Mon269 and Mon270 [41], does not amplify AstV-MLB1. Since it was hypothesized that AstV-MLB1 might represent a whole new group of astroviruses, new primers were designed by Finkbeiner et al. based on an alignment of human astroviruses 1–8 and AstV-MLB1 in order to design new conserved primers that would detect all of these viruses. There are actually very few regions of conservation between AstV-MLB1 and the canonical human astroviruses in which primers can be designed and this is evidenced by the fact that two groups independently designed nearly identical primers [15, 54]. The primers were designed to ORF1b, the polymerase, and yield a ~400 bp amplicon.

These "pan-astroviral" primers were used to screen two cohorts of children with diarrhea: one in Vellore, India, and the other in St. Louis, MO, USA [13, 23]. The Vellore cohort comprised diarrhea samples collected in 2005 and 2006 from regular community sampling of a longitudinal birth cohort and from children hospitalized for acute gastroenteritis. The St. Louis cohort comprised samples sent for bacterial culture to the clinical microbiology laboratory at the St. Louis Children's Hospital between June 2008 and March 2009. The initial intent of the study was to investigate the prevalence of AstV-MLB1, yet surprisingly this study also resulted in the discovery of a number of other novel astroviruses including one that was most similar to AstV-MLB1. It has provisionally been named AstV-MLB2 and shares ~80% amino acid identity to AstV-MLB1. It was detected in 2 out of 416 stools from Vellore and 3 out of 466 stools from St. Louis. AstV-MLB2 was later sequenced and determined to have a genome length of 6,119 nt [23] (Table 7.1).

Astrovirus VA1/Astrovirus HMO-C/HAstV-PS

Five fecal specimens were collected from a 1998 gastroenteritis outbreak at a child care center in Virginia, USA. There were 26 cases of illness in teachers and children reported over a 2- to 3-week period. The samples tested negative for enteric parasites, bacteria, and viruses so they were subjected to high-throughput pyrosequencing and Sanger sequencing by two different groups that reported their findings together [16]. Unique sequences were detected in two of the samples by pryosequencing: 313 sequences in one sample and 1,017 sequences in the other. Astrovirus-like sequences were assembled together into contigs of 6,376 nucleotides (nt) and 6,026 nt from the two samples. These contigs had limited sequence similarity (37–71% aa identity) to proteins from the mink and ovine astroviruses. When the two contigs were assembled together they generated a contig of 6,581 nt in length. The Sanger sequencing that was done simultaneously also resulted in the detection of unique astrovirus-like sequences in the same two samples: 3 in one sample and 69 in the other. Further RT-PCR amplifications and 5′ RACE (rapid amplification of cDNA ends) reactions were used to determine that the full length of the virus is 6,586 nt, excluding the poly-A tail (Table 7.1). This virus has been provisionally named Astrovirus VA1 (AstV-VA1) because it came from an outbreak in Virginia.

AstV-VA1 has three predicted open reading frames and nontranslated regions (NTRs) at both the 5′ and 3′ ends of the genome consistent with the organization of other astroviruses. The lengths of the open reading frames and nontranslated regions are similar to those of mink and ovine astroviruses (Table 7.1). Like most other astroviruses (except TAstV-2 and AstV-MLB1), AstV-VA1 contains the stem-loop II-like motif that is putatively involved in replication. Phylogenetic analysis of the genome confirmed that while this virus is highly divergent from other astroviruses, at the time it was discovered it was most closely related to mink and ovine astroviruses with 61–62% amino acid similarity to them in ORF1b but only 39–42% for the other open reading frames (Fig. 7.1).

In addition to the two groups that both identified AstV-VA1 in association with the Virginia gastroenteritis outbreak, around the same time another group reported the discovery of novel astroviruses in diarrheal stools from Nepal and from stools of kids with nonpoliovirus acute flaccid paralysis in Nigeria and Pakistan [27]. They identified these viruses using similar primers as described above for the discovery of AstV-MLB2. One of the viruses they reported is AstV-VA1. They provisionally gave it the name HMO-C (HMOAstV-C), which stands for *h*uman, *m*ink, and *o*vine-like astrovirus. HMOAstV-C was detected in 3 samples from Nepal out of a total of 328 samples between all of the locations.

Another virus that closely related to AstV-VA1/HMOAstV-C was identified in a 15-year-old patient with X-linked agammaglobulinemia and encephalitis and is described in greater detail later in this chapter [43]. The phylogenetic distance between this virus and AstV-VA1/HMOAstV-C is similar to the distances between human astroviruses 1 and 8, but much less than the distances between any of the other novel astroviruses or between the novel astroviruses and their closest relatives. This raises the question of how to name new viruses and how to define species, which is discussed at the end of this chapter.

Astrovirus HMO-A/Astrovirus VA2

Astrovirus HMO-A (HMOAstV-A), like AstV-VA1/HMOAstV-C, was discovered through PCR screening of a collection of stools from Nepal, Nigeria, and Pakistan [27]. Although seemingly most closely related to AstV-VA1/HMOAstV-C, the phylogenetic distance seems great enough for HMOAstV-A to be considered a different species (Fig. 7.1). It was found in two Nigerian children and one Pakistani child. Pyrosequencing, RT-PCR, and RACE reactions were used to determine the full length genome of HMOAstV-A. Excluding the poly-A tail, it is 6,534 nt long and shares the same organization as other astroviruses (Table 7.1). It contains other common genomic features of astroviruses including a ribosomal frameshift signal between ORF1a and ORF1b, a promoter sequence for putatively initiating ORF2 subgenomic RNA synthesis, a trypsin-like serine protease domain in ORF1a, a proteolytic cleavage site in ORF2, a viral genome-linked protein domain in ORF1a, and the s2m motif in the 3′ NTR.

Just as with AstV-VA1/HMOAstV-C, there were concurrent discoveries of HMOAstV-A. It was also discovered in the same study in which AstV-MLB2 was discovered, involving PCR screening of stools from Vellore, India and St. Louis, MO, USA [13]. It was provisionally named AstV-VA2 in this study and was detected in two Indian stool samples and one American stool sample.

Astrovirus HMO-B/Astrovirus VA3

Astrovirus HMO-B (HMOAstV-B) was the third novel virus to be discovered by PCR in stools from Nepal, Nigeria, and Pakistan [27]. It was found in one sample from Nigeria and one from Pakistan. It is most closely related to the other VA/HMO astroviruses which are all more closely related to mink, ovine, and some of the newly discovered astroviruses in sea lions and bats than to human astroviruses 1–8 or AstV-MLB1 and AstV-MLB2 (Fig. 7.1). The genome was partially sequenced revealing that it has characteristics similar to AstV-VA1/HMOAstV-C and AstV-VA3/HMOAstV-A.

This was the third astrovirus that was discovered both by Finkbeiner et al. and Kapoor et al. in their respective PCR studies [13, 27]. Finkbeiner et al. provisionally named this virus Astrovirus VA3. In their study, it was detected in just one Indian stool making this the least prevalent of the novel VA/HMO astroviruses in both studies.

Epidemiology and Seroprevalence Studies

As described above, these new astroviruses have been found in multiple stool samples collected around the world. Specifically, AstV-MLB1 has been detected in the United States, Mexico, Hong Kong, Egypt, Nigeria, and India [1, 4, 8, 13, 27].

AstV-MLB2 has been detected in diarrhea stools in the United States and India [13]. AstV-VA1/HMOAstV-C has been found in stools from the United States, Nepal, and the Netherlands [16, 27, 43, 47]. AstV-VA2/HMOAStV-A has also been detected in stools from the United States, Egypt, India, and Nigeria [1, 13, 27]. Lastly, AstV-VA3/HMOAstV-B has been detected in stools from India and Pakistan [13, 27]. However, the role of these new viruses in human health and disease is still unknown. While the classic human astroviruses and most of the animal astroviruses are diarrheagenic, it is unknown if the novel human astroviruses are also agents of diarrhea. Ideally, proof that these viruses are a cause of diarrhea would come from human volunteer studies in which subjects would ingest fecal filtrates. However, these types of studies are no longer feasible for novel viruses of unknown pathogenicity. Therefore, proof of disease causality has to now come from indirect studies such as epidemiologic studies.

There has been one case–control study of AstV-MLB1 [22]. This study examined 400 acute diarrhea samples collected from children less than 3 years old in Vellore, India. These diarrhea samples were negative for rotavirus by enzyme immunoassay and PCR, for norovirus by PCR, for bacterial pathogens (*Vibrio cholerae, enteropathogenic Escherichia coli, Salmonella, Shigella, Aeromonas,* and *Plesiomonas*) by culture, biochemical reactions and serogrouping where appropriate and for parasites by routine saline and iodine preparations and modified acid fast stain. Control samples consisted of asymptomatic stool samples obtained from the same child at least 6 weeks prior to the collection of the diarrhea stool. These case and control stools were examined by astrovirus consensus primers which target the RNA polymerase [15]. Samples which were positive by the consensus primers were then screened by RT-PCR using primers specific for AstV-MLB1 and ones to detect the classic human astroviruses. In this study, 4 of the cases and 14 of the controls were positive for AstV-MLB1. Therefore, AstV-MLB1 was less likely to be present in the diarrheal samples than in the asymptomatic samples (OR 0.28, 95% CI 0.09–0.89, $p=0.033$).

Astrovirus VA1/HMO-C was found in the stool of one patient with diarrhea and new onset celiac disease [47]. Stools from four other patients with new onset celiac disease were tested for Ast-VA1 and were negative. Whether AstV-VA1 has a role in the development of celiac disease cannot be determined from this small study, but raises an interesting hypothesis that should be further studied.

To further understand if AstV-VA1/HMOAstV-C is truly a human pathogen or if it is a passenger from ingested food, Burbelo et al. conducted a seroprevalence study [7]. This study used a luciferase immunoprecipitation system to look at human IgG response to a C-terminal capsid fragment of AstV-VA1/HMOAstV-C. This study examined 106 healthy adult blood donors and found 65% to be seropositive for HMOAstV-C/AstV-VA1. Additionally, 103 age-stratified pediatric serum samples were examined. Seropositivity in the 6–12 month, 1–2 year, 2–5 year, and 5–10 year olds was 20%, 23%, 32%, and 36%, respectively. Additionally, it is likely there is transfer of maternal antibodies as 50% (11/22) of the 0–6 month old children showed anti-HMOAstV-C antibody responses. The findings of this study demonstrate that HMOAstV-C/AstV-VA1 is a prevalent infectious agent.

Interestingly, two of the novel astroviruses have been found outside of the enteric system. AstV-MLB2 was found in the plasma of a febrile child by mass sequencing, PCR, and quantitative PCR [23]. This 20-month old had a 3-day history of petechial rash, 1 day of fever, cough, and nasal congestion. The child did not have any vomiting or diarrhea. His evaluation included a normal leukocyte count and a negative blood culture. A nasopharyngeal swab was also negative for respiratory syncytial virus, influenza types A and B, parainfluenza, and adenovirus by fluorescent antibody testing, and culture results were negative for respiratory viruses. Chest X-ray results were interpreted as mild peribronchial thickening, which may represent a viral process. Additionally, plasma or blood of this patient was screened by PCR for the following viruses: adenovirus, enteroviruses, human herpes virus 6 and 7, parvovirus B19, human bocavirus, cytomegalovirus (whole blood), Epstein–Barr virus (whole blood), and JC, BK, WU, and KI polyomaviruses, all of which were negative. Quantitative PCR for AstV-MLB2 showed 4.5×10^5 copies of AstV-MLB2/ml of plasma.

Astrovirus-PS (closely related to AstV-VA1/HMOAstV-C) was found in the brain tissue of a 15-year-old boy with X-linked agammaglobulinemia and encephalitis by deep sequencing, quantitative PCR, and immunofluorescence staining [43]. The cerebral spinal fluid and brain material were negative for the common agents known to cause encephalitis. The histologic findings of the brain material were thought to be consistent with a viral infection.

While these case reports do not prove disease causality, they do suggest that astroviruses may have a role outside of the enteric system. These cases demonstrate a broader distribution of astroviruses in the body than previously recognized.

Nomenclature

With these exciting discoveries of new astroviruses come many challenges. One of these challenges is what to name these new viruses and how to decide if they represent new species. Previously, astroviruses were named for the species they infect (human, turkey, etc.). The classic human species consists of eight closely related serotypes which are named 1–8. In a similar fashion, the HMO astrovirus names are derived from their relationship to human, mink and ovine astroviruses. On the other hand, AstV-MLB1, AstV-VA1, and AstV-PS are named for the site that the sample originated from. A more uniform and instructive naming system needs to be put in place. Kapoor et al. have suggested a naming scheme in which human astroviruses 1–8 are renamed HAstV-1 with a letter (a–h) following the name to denote each serotype and then naming the novel astroviruses described in this chapter HAstV-2–HAstV [27]. This issue of nomenclature is currently being addressed by the International Committee on Taxonomy of Viruses (ICTV) study group for Astroviruses. Additionally, how to define a new species needs to be addressed. There are currently no specific rules in place.

Conclusions

Diarrhea is often not considered a major problem in the developed world, but it still has devastating effects in developing countries. There is a major gap in our understanding of the causes of diarrhea, given that ~40% of all cases are of unknown etiology [25, 33, 48]. While astroviruses are known to cause diarrhea, they have generally been associated with at most 10% of both sporadic and outbreak cases of acute diarrhea. The initial identification of five novel astroviruses in human stool samples over the course of a 2-year period hints that astroviruses may have a larger role in causing human diarrhea than previously believed. Prior to 2008, there were eight known human astroviruses that are all genetically very closely related. It is possible that this paradigm will be upheld for each of the newly described astroviruses in that each one could represent a whole new cluster of astroviruses. As technologies advance and more samples are screened from all over the world, it is likely that other novel astroviruses will be discovered.

The discovery of five novel astroviruses from human samples has nearly doubled the number of known human astroviruses. However, the identification of these novel viruses is surely only the beginning of the story for these viruses. There are many questions that have been raised regarding each virus' tropism, epidemiology, and potential link to disease. Answering these questions will require many years of further investigation, using the available tools of biology and medicine. While all of the recently discovered astroviruses described in this chapter were first identified in stool samples, some have now been detected in other biological specimens or in stools from people without diarrhea. These findings are in no way conclusive, but they do reveal the possibility that astroviruses may be associated with other diseases in humans. The major questions for all of these viruses are (1) Are they truly human pathogens? and (2) If so, what diseases do they cause? These questions are difficult to answer short of doing human infection trials. However, the studies described in this chapter have begun to address these questions.

Historically, the identification of a novel agent in association with a disease has been the rate-limiting step in addressing the question of disease causality. However, a shift has occurred in that the rate-limiting step often now comes after the discovery phase in trying to understand the biological relevance and impact of a newly discovered agent. Determining the role of novel human astroviruses in human health and disease will be an important focus of future efforts as the field of astroviruses moves forward.

References

1. Ahmed SF, Klena SPJD, Pimentel G, Mansour A, Naguib AM, et al. Novel astroviruses in children, Egypt [letter]. Emerg Infect Dis. 2011;17(12):2391–3. doi:10.3201/eid1712.110909.
2. Akihara S, Phan TG, Nguyen TA, Hansman G, Okitsu S, Ushijima H. Existence of multiple outbreaks of viral gastroenteritis among infants in a day care center in Japan. Arch Virol. 2005;150:2061–75.

3. Arnold M, Patton JT, McDonald SM. Culturing, storage, and quantification of rotaviruses. Curr Protoc Microbiol. 2009;Chapter 15:Unit 15C 13.
4. Banyai K, Meleg E, Moschidou P, Martella V. Detection of newly described astrovirus MLB1 in stool samples from children. Emerg Infect Dis. 2010;16:169. author reply 169–170.
5. Belliot G, Laveran H, Monroe SS. Detection and genetic differentiation of human astroviruses: phylogenetic grouping varies by coding region. Arch Virol. 1997;142:1323–34.
6. Brinker JP, Blacklow NR, Herrmann JE. Human astrovirus isolation and propagation in multiple cell lines. Arch Virol. 2000;145:1847–56.
7. Burbelo PD, Ching KH, Esper F, Iadarola MJ, Delwart E, Lipkin WI, et al. Serological studies confirm the novel astrovirus HMOAstrovirus-C as a highly prevalent human infectious agent. PLoS One. 2011;6:e22576.
8. Chu DK, Chin AW, Smith GJ, Chan KH, Guan Y, Peiris JS, et al. Detection of novel astroviruses in urban brown rats and previously known astroviruses in humans. J Gen Virol. 2010;91:2457–62.
9. Clark B, McKendrick M. A review of viral gastroenteritis. Curr Opin Infect Dis. 2004;17:461–9.
10. Davidson GP, Bishop RF, Townley RR, Holmes IH. Importance of a new virus in acute sporadic enteritis in children. Lancet. 1975;1:242–6.
11. Duizer E, Schwab KJ, Neill FH, Atmar RL, Koopmans MP, Estes MK. Laboratory efforts to cultivate noroviruses. J Gen Virol. 2004;85:79–87.
12. Finkbeiner SR, Allred AF, Tarr PI, Klein EJ, Kirkwood CD, Wang D. Metagenomic analysis of human diarrhea: viral detection and discovery. PLoS Pathog. 2008;4:e1000011.
13. Finkbeiner SR, Holtz LR, Jiang Y, Rajendran P, Franz CJ, Zhao G, et al. Human stool contains a previously unrecognized diversity of novel astroviruses. Virol J. 2009;6:161.
14. Finkbeiner SR, Kirkwood CD, Wang D. Complete genome sequence of a highly divergent astrovirus isolated from a child with acute diarrhea. Virol J. 2008;5:117.
15. Finkbeiner SR, Le BM, Holtz LR, Storch GA, Wang D. Detection of newly described astrovirus MLB1 in stool samples from children. Emerg Infect Dis. 2009;15:441–4.
16. Finkbeiner SR, Li Y, Ruone S, Conrardy C, Gregoricus N, Toney D, et al. Identification of a novel astrovirus (astrovirus VA1) associated with an outbreak of acute gastroenteritis. J Virol. 2009;83:10836–9.
17. Fredericks DN, Relman DA. Sequence-based identification of microbial pathogens: a reconsideration of Koch's postulates. Clin Microbiol Rev. 1996;9:18–33.
18. Gary Jr GW, Hierholzer JC, Black RE. Characteristics of noncultivable adenoviruses associated with diarrhea in infants: a new subgroup of human adenoviruses. J Clin Microbiol. 1979;10:96–103.
19. Glass RI, Noel J, Mitchell D, Herrmann JE, Blacklow NR, Pickering LK, et al. The changing epidemiology of astrovirus-associated gastroenteritis: a review. Arch Virol. 1996;12: 287–300.
20. Guix S, Bosch A, Ribes E, Dora L, Dora Martinez L, Pinto RM. Apoptosis in astrovirus-infected CaCo-2 cells. Virology. 2004;319:249–61.
21. Herrmann JE, Hudson RW, Perron-Henry DM, Kurtz JB, Blacklow NR. Antigenic characterization of cell-cultivated astrovirus serotypes and development of astrovirus-specific monoclonal antibodies. J Infect Dis. 1988;158:182–5.
22. Holtz LR, Bauer IK, Rajendran P, Kang G, Wang D. Astrovirus MLB1 is not associated with diarrhea in a cohort of Indian children. PLoS One. 2011;6:e28647.
23. Holtz LR, Wylie KM, Sodergren E, Jiang Y, Franz CJ, Weinstock GM, et al. Astrovirus MLB2 viremia in febrile child. Emerg Infect Dis. 2011;17:2050–2.
24. Jabado OJ, Palacios G, Kapoor V, Hui J, Renwick N, Zhai J, et al. Greene SCPrimer: a rapid comprehensive tool for designing degenerate primers from multiple sequence alignments. Nucleic Acids Res. 2006;34:6605–11.
25. Kapikian AZ. Viral gastroenteritis. JAMA. 1993;269:627–30.
26. Kapikian AZ, Wyatt RG, Dolin R, Thornhill TS, Kalica AR, Chanock RM. Visualization by immune electron microscopy of a 27-nm particle associated with acute infectious nonbacterial gastroenteritis. J Virol. 1972;10:1075–81.

27. Kapoor A, Li L, Victoria J, Oderinde B, Mason C, Pandey P, et al. Multiple novel astrovirus species in human stool. J Gen Virol. 2009;90:2965–72.
28. Kiang D, Matsui SM. Proteolytic processing of a human astrovirus nonstructural protein. J Gen Virol. 2002;83:25–34.
29. Kurtz JB, Lee TW. Human astrovirus serotypes. Lancet. 1984;2:1405.
30. Larkin MA, Blackshields G, Brown NP, Chenna R, McGettigan PA, McWilliam H, et al. Clustal W and Clustal X version 2.0. Bioinformatics. 2007;23:2947–8.
31. Lee TW, Kurtz JB. Human astrovirus serotypes. J Hyg (Lond). 1982;89:539–40.
32. Lee TW, Kurtz JB. Prevalence of human astrovirus serotypes in the Oxford region 1976-92, with evidence for two new serotypes. Epidemiol Infect. 1994;112:187–93.
33. Lyman WH, Walsh JF, Kotch JB, Weber DJ, Gunn E, Vinje J. Prospective study of etiologic agents of acute gastroenteritis outbreaks in child care centers. J Pediatr. 2009;154:253–7.
34. Madeley CR, Cosgrove BP. Letter: 28 nm particles in faeces in infantile gastroenteritis. Lancet. 1975;2:451–2.
35. Mardis ER. The impact of next-generation sequencing technology on genetics. Trends Genet. 2008;24:133–41.
36. Matsui SM, Kim JP, Greenberg HB, Young LM, Smith LS, Lewis TL, et al. Cloning and characterization of human astrovirus immunoreactive epitopes. J Virol. 1993;67:1712–5.
37. Mendez E, Arias CF. Astroviruses. In: Knipe DM, Howley PM, editors. Fields virology, vol. 1. 5th ed. Philadelphia: Lippincott Williams & Wilkins; 2007.
38. Mendez E, Salas-Ocampo E, Arias CF. Caspases mediate processing of the capsid precursor and cell release of human astroviruses. J Virol. 2004;78:8601–8.
39. Monceyron C, Grinde B, Jonassen TO. Molecular characterisation of the 3′-end of the astrovirus genome. Arch Virol. 1997;142:699–706.
40. Nichol ST, Spiropoulou CF, Morzunov S, Rollin PE, Ksiazek TG, Feldmann H, et al. Genetic identification of a hantavirus associated with an outbreak of acute respiratory illness. Science. 1993;262:914–7.
41. Noel JS, Lee TW, Kurtz JB, Glass RI, Monroe SS. Typing of human astroviruses from clinical isolates by enzyme immunoassay and nucleotide sequencing. J Clin Microbiol. 1995;33:797–801.
42. O'Ryan M, Prado V, Pickering LK. A millennium update on pediatric diarrheal illness in the developing world. Semin Pediatr Infect Dis. 2005;16:125–36.
43. Quan PL, Wagner TA, Briese T, Torgerson TR, Hornig M, Tashmukhamedova A, et al. Astrovirus encephalitis in boy with X-linked agammaglobulinemia. Emerg Infect Dis. 2010;16:918–25.
44. Ronaghi M, Karamohamed S, Pettersson B, Uhlen M, Nyren P. Real-time DNA sequencing using detection of pyrophosphate release. Anal Biochem. 1996;242:84–9.
45. Rose TM, Schultz ER, Henikoff JG, Pietrokovski S, McCallum CM, Henikoff S. Consensus-degenerate hybrid oligonucleotide primers for amplification of distantly related sequences. Nucleic Acids Res. 1998;26:1628–35.
46. Sanger F, Nicklen S, Coulson AR. DNA sequencing with chain-terminating inhibitors. Proc Natl Acad Sci USA. 1977;74:5463–7.
47. Smits SL, van Leeuwen M, van der Eijk AA, Fraaij PL, Escher JC, Simon JH, et al. Human astrovirus infection in a patient with new-onset celiac disease. J Clin Microbiol. 2010;48:3416–8.
48. Svraka S, Duizer E, Vennema H, de Bruin E, van der Veer B, Dorresteijn B, et al. Etiological role of viruses in outbreaks of acute gastroenteritis in The Netherlands from 1994 through 2005. J Clin Microbiol. 2007;45:1389–94.
49. Tiemessen CT, Kidd AH. The subgroup F adenoviruses. J Gen Virol. 1995;76(Pt 3):481–97.
50. Walter JE, Briggs J, Guerrero ML, Matson DO, Pickering LK, Ruiz-Palacios G, et al. Molecular characterization of a novel recombinant strain of human astrovirus associated with gastroenteritis in children. Arch Virol. 2001;146:2357–67.
51. Walter JE, Mitchell DK. Astrovirus infection in children. Curr Opin Infect Dis. 2003;16:247–53.
52. Wilhelmi I, Roman E, Sanchez-Fauquier A. Viruses causing gastroenteritis. Clin Microbiol Infect. 2003;9:247–62.

53. Woo PC, Lau SK, Chu CM, Chan KH, Tsoi HW, Huang Y, et al. Characterization and complete genome sequence of a novel coronavirus, coronavirus HKU1, from patients with pneumonia. J Virol. 2005;79:884–95.
54. Zhu HC, Chu DK, Liu W, Dong BQ, Zhang SY, Zhang JX, et al. Detection of diverse astroviruses from bats in China. J Gen Virol. 2009;90:883–7.

Chapter 8
Astrovirus Infections in Animal Mammalian Species

Giovanni Cattoli, Daniel Ka Wing Chu, and Malik Peiris

Abstract The first reports of astroviruses in animals date back to the end of the 1970s, when infections in mammals such as lambs and calves suffering from diarrhea were reported for the first time. Since then, several mammalian species have been shown to be susceptible to astroviruses which appear to be genetically diverse and to have acquired host-specificity. To date, astroviruses have been detected in 16 different orders or species of mammals in addition to humans, and signs of infection range from unapparent infection or very mild disease to diarrhea, lethargy, and anorexia, mainly observed in young individuals. This chapter describes those astroviruses detected in nonhuman mammalian species worldwide, as well as their molecular and phenotypic characteristics and their role in diseases. The capacity of these viruses to cross-species barriers and their subsequent adaptation to novel hosts is also highlighted.

Introduction

Following the first description of astroviruses in human stool in 1975 [1, 27], astrovirus-like particles were described and reported in domesticated animals. The first reports derived from infections in mammals such as lambs and calves

---

G. Cattoli, D.V.M., Ph.D. (✉)
Department of Research and Innovation, Istituto Zooprofilattico Sperimentale delle Venezie, Viale dell'Università 10, Legnaro 35020, Italy
e-mail: gcattoli@izsvenezie.it

D.K.W. Chu, Ph.D. • M. Peiris, M.B.B.S., D.Phil., M.R.C. Path.
Centre for Influenza Research and School of Public Health, The University of Hong Kong, 5/F., William MW Mong Building, Li Ka Shing Faculty of Medicine, Pokfulam, People's Republic of China
e-mail: dkwchu@gmail.com; malik@hku.hk

S. Schultz-Cherry (ed.), *Astrovirus Research: Essential Ideas, Everyday Impacts, Future Directions*, DOI 10.1007/978-1-4614-4735-1_8,

suffering from diarrhea [40, 48]. Indeed, between the end of the 1970s and the first half of the 1980s several animal species were found positive for astrovirus and many novel animal astroviruses were described probably as a consequence of the description of the discovery of astrovirus infection in humans few years before. Over the next 20 years, there was a hiatus of information about the occurrence of novel astroviruses or the clinical manifestations associated with known astroviruses in mammalian species. The lack of a clear association of astrovirus detection in clinical specimens with signs of disease in animals, together with the limited sensitivity and availability of laboratory tests (i.e., electron microscopy and virus isolation) contributed to this gap in information. The advent of the genomic and metagenomic era in the recent period appears to have revitalized investigations in this field and there is now new interest in the circulation of astroviruses in animal and human populations, with an increasing number of reports concerning new host species infected and novel astroviruses described in the last ten years.

Astrovirus Infections in Domestic Ruminants (Ovine and Bovine)

Ovine Astrovirus

Ovine astrovirus (OAstV) replicates in the absorptive epithelial cells of the small intestine and induces mild diarrhea in gnotobiotic lambs [13, 41]. The genome of OastV isolated in 1977 has been completely sequenced in more recent years [22], although information on its capsid protein sequence were published a few years before [21]. Results from these papers indicated that human, feline, and porcine astroviruses were most closely related to each other than to OAstV. However, as with most of the known astroviruses, OAstV harbors a RNA motif in the 3′-end of the genome common to distinct and genetically different virus families (i.e., *Coronaviruses* and *Picornaviruses*) and is supposed to derive from a recombination event between these viruses [20]. Genetic sequencing of astroviruses recently led to the detection of viruses in human stool that are genetically closely related to OAstV and mink astrovirus (MiAstV), and has revealed the possibility that these viruses have a common ancestor, as previously suggested also for the HAstVs, feline astroviruses (FAstVs), and porcine astroviruses (PAstVs) [11, 20, 23, 35]. Despite the detailed genetic studies undertaken on the genome of this virus, no further clinical or epidemiological reports indicating its occurrence and circulation in animals have appeared in the last 35 years. As a consequence, the epidemiology as well as the genetic and antigenic diversity of OAstV is poorly defined and only the complete genome sequence of the original isolate is available in the public genetic database (GenBank, http://www.ncbi.nlm.nih.gov).

Bovine Astrovirus

One year after the detection of astrovirus in lambs, the first isolate of bovine astrovirus (BAstV) was reported in England [48]. Astrovirus was considered to be avirulent as experimentally infected gnotobiotic calves remained clinically normal, although pathological studies on infected calves were not performed at that time. In Florida in1984, a bovine enteric virus antigenically related to the UK strain of BAstV was shown to cause infection and cytopathology of the M cells of the dome epithelium covering Peyer's patches of the ileum of gnotobiotic calves. Similar to the England strain, these animals remained clinically normal during the course of the infection and only a change in the color of the feces from brown to yellow was noticed [49]. In 1985, a further isolate of BAstV was obtained from a colostrum-deprived calf which developed diarrhea [50]. Serotyping of the strains from England and USA based on neutralization and immunofluorescence assays indicated that these viruses were antigenically related although likely belonging to different serotypes. The antigenic diversity detected in the earlier studies is also reflected in the more recently described genetic diversity for BAstV. Tse et al. [44] detected sequences from two different BAstVs coinfecting 2 out of 5 positive specimens [44]. The phylogenetic analyses support classification of BAstVs and the newly discovered astroviruses in roe deer (CCAstV) under the proposed "genocluster GI" of the *Mamastrovirus* [4, 38, 44].

Originally, a serologic study conducted on cattle in the United States led to the detection of a 30% seropositivity for astrovirus [49] but the limited clinical significance initially attributed to BAstV infection in cattle probably explains why there has been a lack of further reports for more than 25 years, until 2011 [44]. Although a previous study suggested that BAstV could be excreted by 60–100% of calves on farms [5], a much lower rate of isolation (2.6%) was found by analyzing 1,060 field samples of diarrhea in calves aged from 3 days to 3 weeks of age [50]. A very similar detection rate (2.4%) was revealed by analyzing 209 rectal swabs collected from asymptomatic adult cattle by RT-PCR [44]. It was originally concluded that in natural conditions, BAstV did not seem to be directly associated with a severe diarrheic disease in calves [5, 44, 48, 49] and few controversial data are available on the prevalence of this infection. For this reason, it has recently been recommended that a large-scale surveillance study for BAstV should be performed in commercially farmed cattle to re-evaluate the circulation and the epidemiology of this virus.

Astrovirus Infections in Other Ruminants

Soon after the reports of astrovirus infections in lambs and calves, astrovirus-like particles were detected by electron microscopy in the diarrheic feces of red deer reared in captivity in the UK [45]. Attempts to propagate the viruses by serial passages on bovine embryonic kidney cells failed but seroconversion was detected in convalescent sera collected 2 weeks after virus shedding. Following this report, no more cases of astrovirus infections in red deer have been published and the role of these viruses in the enteric outbreak described in the UK remains uncertain. However, an outbreak of enteric disease recently occurred in a dense wild population of European roe deer (*Capreolus capreolus*) in Denmark at the beginning of 2010. Virological investigations led to the identification and characterization of novel astroviruses (CcAstV) in the absence of evidence of other viral, bacterial, or parasitic pathogens [38]. In the original study, two CcAstV detected in two distinct samples revealed a relatively low amino acid sequence identity (84.9%) in the region encoded by the RdRP gene. As the genetic distance between the two CcAstV is at least as large as that between the eight human astrovirus serotypes, the authors concluded that these novel astroviruses may constitute two different serotypes [38]. The viruses responsible for the infection was shown to genetically belong to the genus *Mamastrovirus* and are related to BAstV, suggesting that this virus could cross the species barrier to infect both cattle and roe deer [38, 44].

Astrovirus Infections in Carnivores

Feline Astrovirus

After the first clinical report in 1981, astrovirus-like particles have been described in the feces of domestic cats in Australia, England, Germany, New Zealand, and the USA [15, 17, 18, 29, 36]. The first report originating from the USA described the occurrence of severe and prolonged enteric signs with green and watery diarrhea for about 2 weeks, poor body condition and dehydration in a female 4-month-old kitten [18]. Typical astrovirus-like particles were detected in feces by electron microscopy but virus isolation attempts in primary feline kidney cells and fetal rhesus monkey kidney cell line failed. A few years later, a naturally acquired astrovirus infection was described in a 1-year-old domestic cat in England that was presented to the veterinary clinic with a history of intermittent vomiting and diarrhea [15]. No parasites or significant bacterial pathogens were detected in the feces but the presence of a large number of astrovirus-like particles was documented. Fecal filtrates containing astrovirus-like particles were administered to 14-week-old specific pathogen-free kittens causing two episodes of pyrexia and mild diarrhea [15]. Inoculated SPF kittens had intermittent excretion of virus between day 4 and day 6 postinfection. Viruses from feces of SPF kittens were isolated on feline embryo cells incubated in the presence of 10–50 μg of trypsin. Under experimental conditions, events of pyrexia coincided with episodes of virus shedding and seroconversion was observed in the experimentally infected animals by immunofluorescence assay. No cross reactivity between feline and human astroviruses was observed by immune electron microscopy [15]. Based on the report from England, astrovirus-like infection in kittens appeared to be associated to clinical disease. However in a 16-month study conducted in Australia, the occurrence of this viral infection was not related to the age of cats and to the presence of diarrhea [29]. Interestingly, in this study, a human serum was found to react against a cat astrovirus by immune electron microscopy, implying similarity in antigenic regions between human and FeAstV [29]. A later report from New Zealand described the presence of an astrovirus-like particle in the diarrheic feces of two young kittens [36]. Since 1993, no further reports of astrovirus infections in cats have been reported and the role of these viruses in feline enteritis and their epidemiology remains obscure. Furthermore, to date, only partial sequences of the ORF1b and of the ORF2 of these viruses have been deposited in public databases and feline astrovirus diversity has not yet been explored. Interestingly, a study on both astrovirus genome sequences and host species evolution has led to the hypothesis of a likely cross-species transmission from pigs to cats and further from cats to humans, possibly involving intermediate species [25].

In 2009, a novel astrovirus was identified in captive cheetahs (*Acinonyx jubatus*) showing clinical signs of gastroenteric disease, such as lethargy, anorexia, watery diarrhea, and regurgitation [2]. The virus was shown to be related to the FeAstV in ORF2 (88.6%, 97.1% nucleotide and amino acid identity, respectively). Currently, it is unclear whether the cheetah astrovirus crossed the species barrier from domestic cats or if it was established in cheetahs independently.

Canine Astrovirus

Astrovirus-like particles have been reported in dogs with and without diarrhea since 1980 [28, 46, 47]. However, these reports were based on electron microscopy descriptions only, thus the correct and complete identification of the small virus particles detected was not possible. The occurrence of astrovirus infection in dogs was only recently confirmed by RT-PCR detection and further genetic characterization by sequencing [43, 53]. In 2011, a canine astrovirus (CaAstV) has also been successfully isolated in cell culture for the first time [30]. To date, CaAstV or astrovirus-like particles in dogs have been reported in the USA, Germany, Austria, Italy, China, and France, suggesting the worldwide circulation of these viruses [14, 28, 43, 46, 47, 53]. The clinical significance of CaAstV infections in dogs requires further investigation; however, CaAstV was detected from symptomatic puppies occasionally suffering from diarrhea in association with other enteric viruses, such as rotaviruses [43] and corona- and/or parvo-like viruses [30, 47]. The clinical sign most commonly associated with the presence of astrovirus in puppies is a self-limiting watery diarrhea. However, based on RT-PCR virus shedding can persist long after the termination of the clinical signs [30]. A recent study on the prevalence of CaAstV in Shanghai, China, suggested that 12% (22/183) of the puppies showing clinical signs of diarrhea were positive for astrovirus by RT-PCR as compared to none of the 138 healthy controls [53]. In a recent study conducted in Italy, 24.5% of 110 stool samples collected from symptomatic dogs and 9.3% of 75 from asymptomatic animals surveyed tested positive for the presence of CaAstV RNA [30]. In the same study, serologic assay indicated that 59% of 54 dogs aged >3 months surveyed in Italy presented specific CaAstV antibodies [30]. A similar prevalence based on viral RNA detection was found in France where 20.9% (66/316) of the puppies in 42% (14/33) in breeding kennels surveyed were CaAstV-positive by RT-PCR [14]. In the same report, the authors noted that puppies less than 7 week-old were especially susceptible to CaAstV infection [14]. These data indicate that CaAstV is ubiquitous in the canine populations in several parts of the world.

Genetic analysis of ORF2 and partial ORF1b identified the virus as a member of the genus *Mamastrovirus*, genocluster GI comprising "classic" human, porcine, and FeAstVs but genetically well distinguishable from the other *Mamastroviruses* [4, 43]. Among available sequences, the closest similarity was found with HAstV-7 and HAstV-1 (22% and 59.4% nucleotide similarity for ORF2 and partial ORF1b sequences, respectively). Thus, genospecies GI.E has been further proposed as being a new species [4]. Based on the genetic data reported from Italy and China, the genetic variability of CaAstV also appears to be relevant within the same country and perhaps indicative of the existence of multiple antigenic types [30, 43, 53].

Replication in vitro of CaAstV was successfully replicated on the MDCK cell line supplemented with 10 μg/ml of trypsin after three passages. Enlargement and detachment of the cells and the appearance of fine cytoplasmatic granules were the main features of the cytopathic effect [30].

Mink Astrovirus

A preweaning syndrome has been well described in mink farms in Denmark since 1954 [42]. This syndrome usually affects minks 1–4 weeks of age and mink farmers refer to it as "sticky," "greasy," or "wet" kits, due to the hypersecretion from apocrine glands in the dorsal neck region resulting in soiling of the neck and back [39], together with diarrhea and dehydration at various degrees [16]. The postmortem lesions described in kits dying from this syndrome are catarrhal enteritis with villous atrophy and desquamation of epithelial cells at the tip of intestinal villi [16, 19]. Many causative factors have been associated with the occurrence of this syndrome, including diet, bacterial, and viral pathogens [10]. In 2002, astrovirus infection was epidemiologically linked to the preweaning syndrome in minks, although its definitive etiology remained obscure [10]. In 2003, the astrovirus associated with this syndrome was entirely sequenced and characterized as a novel astrovirus with less than 67% similarity at the nucleotide level with the closest related OAstV, thus called MiAstV [32, 33]. Interestingly, sequence analysis of MiAstVs from geographically distinct Swedish and Danish farms indicated that they were conserved genetically, although this observation was mainly based on a short fragment of the ORF1b and geographical clusters were revealed [32, 33]. This appears to be in contrast with the data collected for astroviruses infecting other species, such as swine, bats or turkeys, indicating a high degree of genetic variability [6, 26, 34, 52]. More recently, a novel astrovirus has been discovered in brain tissue of mink developing the so-called shaking mink syndrome (SMS), a transmissible neurological syndrome described for the first time in farmed minks of Denmark, followed by detection in both Sweden and Finland. The main clinical signs are shaking, staggering gait, and ataxia, with no evident gross lesions present postmortem. Nonsuppurative encephalomyelitis is a common lesion revealed by histopathology [3, 12]. Blomstrom et al. [3] have identified a unique astrovirus associated with SMS. The astrovirus associated with SMS and found in the brain tissue of minks was very similar to MiAstV in the ORF1a and ORF1b regions (88% and 98% homology, respectively) and a lower identity for ORF2 (67% and 59% homology for nucleotide and amino acid sequences, respectively) [3]. Interestingly, in 2007, a lethal case of human encephalitis occurred in a boy with X-linked agammaglobulinemia and an astrovirus phylogenetically distinct from HAstVs but related to MiAstV was detected in the brain tissue. Although the definitive source of infection remained unknown, it should be noted that the patient's residence was located in proximity to a mink farm [35].

Astrovirus Infections in Sea Mammals and Other Species

Five genetically distinct astroviruses have recently been found in the fecal samples of three species of marine mammal: a California sea lion (*Zalophus californianus*) (CslAstV-1, CslAstV-2, CslAstV-3); a Steller's sea lion (*Eumetopias jubatus*) (SslAstV); and a bottlenose dolphin (*Tursiops truncates*) (BdAstV, 37). As for astrovirus infections in other species, the clinical significance of infection in sea mammals is unclear. In two cases (California sea lion and Steller sea lion) astroviruses were detected in young animals. The free-ranging California sea lion tested positive for astrovirus suffered from diarrhea but also had salmonella [37]. Genetic analyses of ORF1b and ORF2 indicate that these viruses do not cluster in a monophyletic group but across the *Mamastrovirus* tree, as confirmed by the proposal of reclassification submitted to the ICTV. Recombination analyses suggested that a recombination event may have occurred between a human strain (HAstV-4) and a CslAstV-2, possibly resulting in CslAstV-3 [37].

At the time of writing, the most recently discovered mammalian astrovirus infection is in rabbits. A recent investigation conducted in Italy on farmed rabbits suffering from enterocolitis revealed a novel astrovirus distantly related to other mammalian astroviruses, based on the ORF2 encoding for the capsid protein [31]. According to this study, astroviruses were detected in the intestinal content or feces of both symptomatic and asymptomatic rabbits. Considering the clinical relevance in rabbit farming of enteritis complex syndrome, this finding certainly deserves further investigations to assess the real impact and role of astrovirus infections in this species.

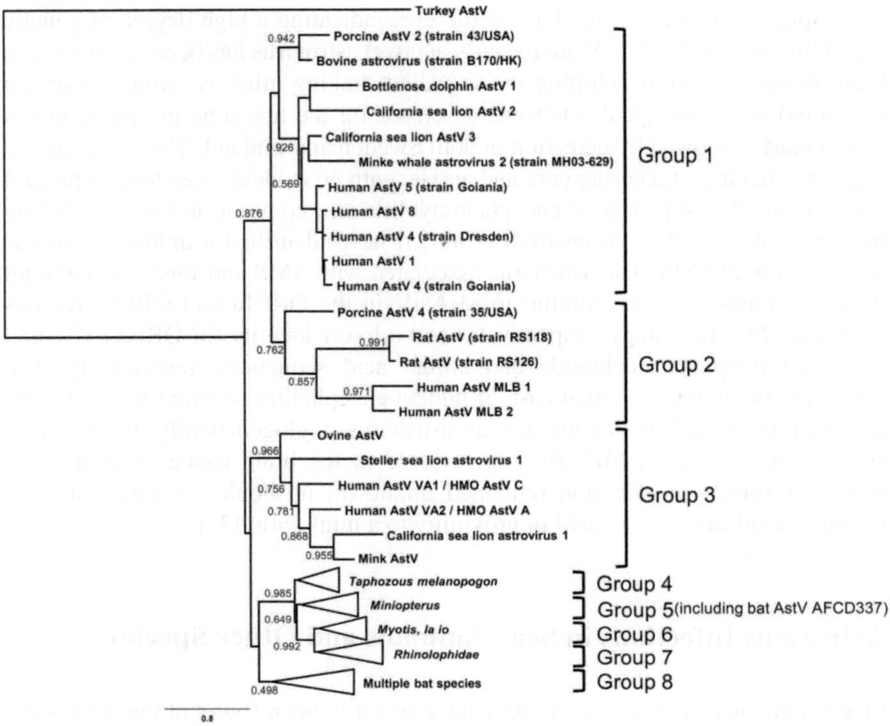

Fig. 8.1 Phylogenetic analysis of the RdRp gene (422 nt) of mamastroviruses using PhyML with alrt (approximate likelihood ratio test) values indicated. Eight groups of mamastroviruses are denoted. A turkey astrovirus was included as an outgroup in this analysis

Astroviruses in Rats

Fecal samples from *Rattus norvegicus* (brown rats, $n = 371$) and *Rattus rattus* (black rats, $n = 70$) captured in an urban area of Hong Kong were tested for astroviruses using broad-range primers designed to be conserved across all available astroviruses genetic sequences. While none of the *R. rattus* ($n = 70$) tested were positive for astroviruses, 6 (1.6%) of 371 *R. norvegicus* specimens had astroviruses. Attempts at culturing these viruses in BHK, Caco-2, Vero or HEK293 cells, with or without exogenously added trypsin was unsuccessful.

Phylogenetic analysis of partial genome sequences of the RdRp gene of these rat astroviruses revealed that they cluster within a separate group distinct from all other astroviruses and form two clades represented by AstVRS118 and AstV RS126, respectively. The rat astroviruses have a sister clade relationship to human astroviruses MLB1 and MLB2 (Fig. 8.1). The phylogenetic analysis suggests that the rat and human MLB1 and 2 viruses appear to share a common ancestor virus, the mean date of the most recent common ancestor virus was estimated to have

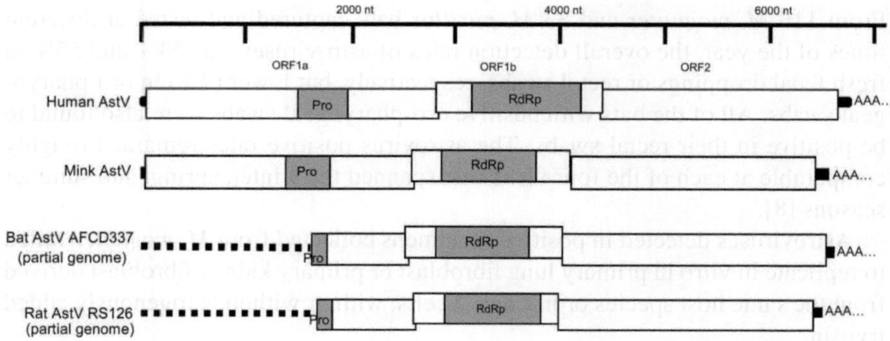

Fig. 8.2 Schematic comparing the organization of typical human, mink, bat, and rat astrovirus genomes. The partial genome structure of the novel bat and rat astroviruses showed a typical astrovirus genome organization

occurred around 1054 A.D. (95% highest posterior densities 837–1827 A.D.) *R. norvegicus* is known to be a peri-domestic rodent species of human settlements for centuries and it is plausible that human AstV MLB and these rat astroviruses arose from a common ancestor centuries ago. The analysis of dN/dS ratios of ORF1b sequences does not show increased positive selection for either rat astrovirus or MLB viruses suggesting that each has been well adapted in the respective hosts over long periods of time [9].

Two-thirds of the genome of these two representative viruses was genetically sequenced, including part of ORF1a and the complete ORF1b and ORF2 genes and the 3′ UTR. A serine protease and RdRp gene can be identified in the ORF1a and ORF1b, respectively, from both rat astrovirus sequences. A heptameric slippery sequence (AAAAAAC) can be identified in the putative ORF1a/1b junction. In rat astrovirus, as with bat astrovirus AFCD337 and MLB virus, the conserved nucleotide sequence of a stem-loop II-like motif which is commonly found in many other astroviruses could not be found (Fig. 8.2).

Rat astroviruses may potentially provide a physiologically relevant small mammal model of astrovirus infections that may help understanding the pathogenesis of astrovirus infections. Similarly, CaAstVs have also been cultured in vitro [30] and could provide an alternative animal model.

Astroviruses in Chiroptera

Following the detection of diverse group 1 and group 2 coronaviruses, including the putative precursors of SARS coronavirus, in insectivorous bats, there has been increased interest in the virus ecology of these species. Astroviruses were initially detected by consensus PCR methods in rectal swabs of apparently healthy *Miniopterus magnater* and *Miniopterus pusillus* bats captured in Hong Kong.

From 116 *M. magnater* and 33 *M. pusillus* bats captured and tested at different times of the year, the overall detection rates of astroviruses was 54% and 55% in fresh fecal droppings or rectal swabs, respectively, but lower (7%) in oro-pharyngeal swabs. All of the bats with positive oro-pharyngeal swabs were also found to be positive in their rectal swabs. The astrovirus positive rates remained roughly comparable at each of the four visits that spanned the winter, spring, and summer seasons [8].

Astroviruses detected in positive specimens collected from *M. magnater* failed to replicate in vitro in primary lung fibroblast or primary kidney fibroblast derived from the same host species or in Caco-2 cells, with or without exogenously added trypsin.

Approximately three-fourths of the genome of a bat astrovirus (AFCD337) detected in the rectal swab of a *M. pusillus* bat has been sequenced, including part of the ORF1a, and complete ORF1b, ORF2 and the 3′ untranslated region and poly-A tail at the 3′end of the genome (Fig. 8.2). These studies show that bat astroviruses are distinct from other known mammalian astroviruses (Fig. 8.1). In the ORF1b and ORF2 regions, batAstV AFCD337 has <55% and <27% amino acid similarity, respectively, to other known mammalian astroviruses. Characteristic features of BatAstV AFCD337 include a protease motif in ORF1a, an astrovirus "slippery sequence" (AAAAAC) at the junction between ORF1a and ORF1b which is required for inducing a ribosomal shift event, a RdRp motif in ORF1b and a conserved sequence at the end of ORF1b [8]. A conserved stem-loop structure that is predicted at the 3′-end of the genomic RNA seen in human, ovine, porcine, and turkey astroviruses was not found in BatAst AFCD337 [8, 20, 22].

The putative ORF2 of AFCD337 has a size of 2,553 nt and bat viruses appear to have the largest astrovirus capsid gene known to date. As with other astroviruses, the N-terminal half of ORF2 which is believed to be the core assembly domain of the viral capsid [24] is more conserved. Compared to an overall amino acid similarity of <27% for the ORF2 region overall, the amino acid sequence similarities of the N-terminal half of ORF2 to HAstV-1 Oxford, OAstV, and MAstV being 36.3%, 45%, and 39.5%, respectively. Phylogenetic analysis of ORF1a (partial), ORF1b, and ORF2 confirms that BatAst AFCD337 is a novel astrovirus (Fig. 8.1).

Representative insectivorous bats of families Rhinolophidae (12 species tested), Vespertilionidae (12 species), Emballonuridae (*Taphazous melnopogon*), Megadermatidae (*Megaderma lyra*), and Hipposideridae (2 species) and fruit bats of the family Pteropodidae (*Rousettus leschenaultia*) captured or collected in southern China have been tested [8, 51, 52]. The insectivorous bats had relatively high detection rates of astroviruses. In contrast, the only fruit bat species tested (*R. leschenaultia)* had low rates of astroviruses detected. While one astrovirus detected from *R. leschenaultia* in China clustered phylogenetically with viruses from Rhinolophidae [51], one other virus from this species was phylogenetically distinct from all astroviruses detected from insectivorous bats (unpublished data).

The RdRp gene (422 nt) of these bat astroviruses detected so far has been sequenced to compare the evolutionary diversity of these viruses in comparison to other mammalian astroviruses. All bat astroviruses clustered within the genus Mamastrovirus. Within this genus, eight monophyletic groups can be identified. Group 1 contains the typical human astroviruses and those from porcine, bovine, and sea mammal origin. Group 2 contains human atypical AstV MLB1 and MLB2, rat, and a porcine astrovirus. Group 3 comprises mink, ovine, and atypical human astroviruses VA1/2/3 (also alternatively termed HMO astroviruses A/B/C). Groups 4, 5, 6, and 7 were monophyletic clusters of viruses from different families of bats, viz., Taphozous, Miniopterus, Myotis, and Rhinolophidae, respectively. Group 8 has weak statistical support and has viruses from diverse bat species including those from Taphozous, Miniopterus, and Myotis bats (Fig. 8.1).

Two features of the ecology and phylogeny of bat astroviruses are unusual. The virus detection rates are high from apparently healthy animals and there is great genetic diversity even within bats of the same species at the same habitat at one sampling occasion (e.g., in *Miniopterus* bats within a single bat cave in Hong Kong). Astrovirus infections are typically acute short lasted infections and while virus detection rates >20% may be found in the context of outbreaks of infection, to find such (or even higher) detection rates repeatedly within a single species within one habitat is surprising. It is even more surprising to find high genetic diversity from animals within the same habitat at one sampling occasion, so that almost every animal has a genetically unique virus. The RdRp gene sequences of 50 viruses from one monophyletic group of astroviruses (from *Miniopterus* bats) are shown (Fig. 8.3). Pairwise genetic distances of the RdRp gene of these monophyletic viruses were estimated for all virus pairs and we found that 96.2% of these pair-wise distances were larger than or equal to 0.18 nucleotide substitution per site. In contrast, the pair-wise genetic distance analysis for all available typical human astroviruses worldwide was never larger than 0.17 nucleotide substitution per site. Thus >96% of all the pairs of bat astroviruses in this monophyletic group in a single habitat have diversity larger than that of typical human astroviruses considered globally.

In 50 astroviruses detected from this population of *M. magnater* bats, 45 viruses had genetic distances of at least 0.18 nucleotide substitutions per sites in the RdRp gene to their closest partner. Assuming an RNA virus mutation rate of 10^{-3} per year, 0.18 substitutions per site represents separation of >180 years for each virus pair found within a single habitat, i.e., the most recent common ancestor for any two viruses would be at least 90 years. The observation that almost each virus detected in Miniopterus bats even in one sampling occasion was genetically unique is reminiscent of the findings on bat coronaviruses in this same population [7] and raises the question whether astroviruses may persistently infect bats and whether bats are unusual in their virus–host interactions with these viruses. Longitudinal studies to re-sample tagged bats may be required to address these questions. An alternative explanation that these viruses found in bats are insect viruses ingested by these insectivorous bats is implausible given the close phylogeny with other mammalian viruses and also the clear clustering of viruses within bat families.

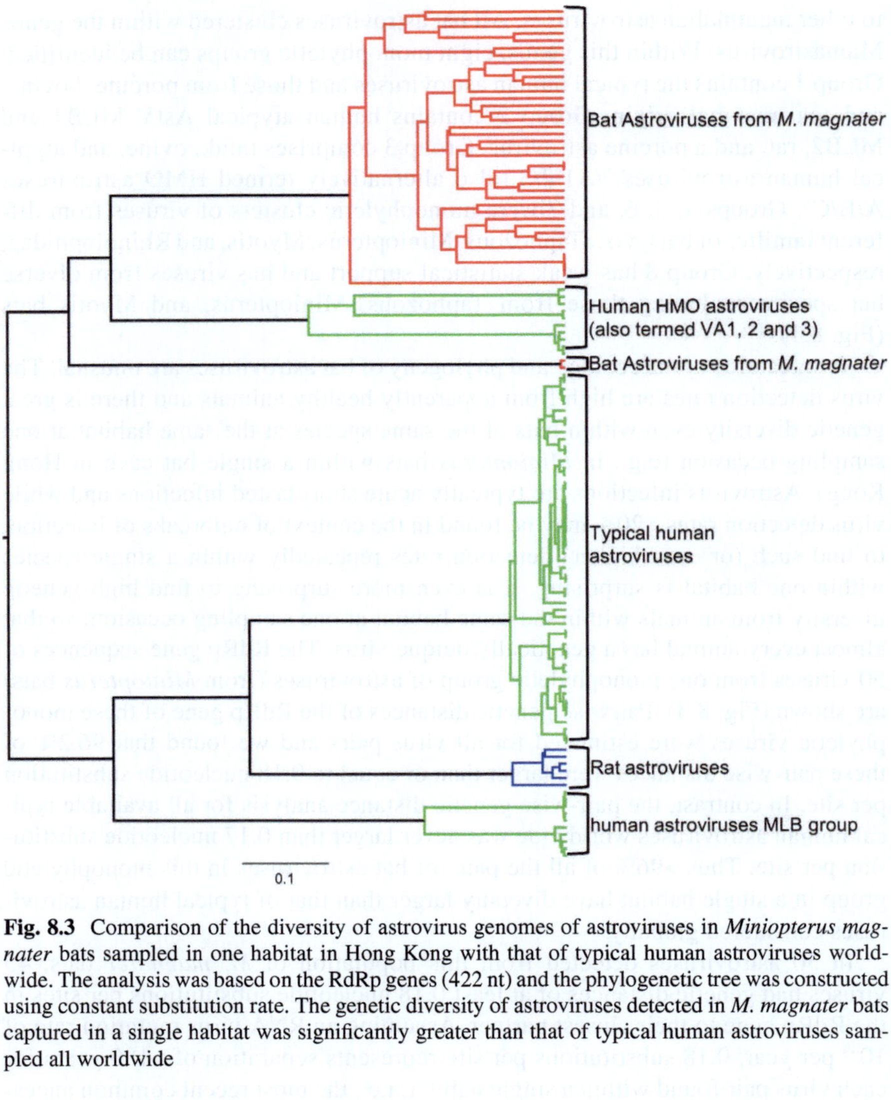

Fig. 8.3 Comparison of the diversity of astrovirus genomes of astroviruses in *Miniopterus magnater* bats sampled in one habitat in Hong Kong with that of typical human astroviruses worldwide. The analysis was based on the RdRp genes (422 nt) and the phylogenetic tree was constructed using constant substitution rate. The genetic diversity of astroviruses detected in *M. magnater* bats captured in a single habitat was significantly greater than that of typical human astroviruses sampled all worldwide

Nomenclature of Astroviruses

The International Committee for Taxonomy of Viruses has estimated within species and between species amino acid sequence diversity in the capsid gene of ≤0.312 and ≥0.378, respectively. The capsid gene is generally more genetically diverse than the RdRp gene. Thus, if these criteria are directly applied to astroviruses in bat populations, even within a single bat species, viruses from most individual bats within one bat species would qualify as a different virus species.

This issue would have to be considered with some flexibility, especially in regard to bat astroviruses. Given that different bat species from different regions of China group together in monophyletic clades, even though their within group diversity is large, may have to be considered as one species.

References

1. Appleton H, Higgins PG. Letter: Viruses and gastroenteritis in infants. Lancet. 1975; 305 (7919):1297.
2. Atkins A, Wellehan Jr JF, Childress AL, Archer LL, Fraser WA, Citino SB. Characterization of an outbreak of astroviral diarrhea in a group of cheetahs (Acinonyx jubatus). Vet Microbiol. 2009;136(1–2):160–5.
3. Blomstrom AL, Widen F, Hammer AS, Belak S, Berg M. Detection of a novel astrovirus in brain tissue of mink suffering from shaking mink syndrome by use of viral metagenomics. J Clin Microbiol. 2010;48(12):4392–6.
4. Bosch A, Guix S, Krishna NK, Méndez E, Monroe SS, Pantin-Jackwood M, Schultz-Cherry S. Nineteen new species in the genus Mamastrovirus in the Astroviridae family. ICTV 2010.018. aV. 2010b.
5. Bridger JC, Hall GA, Brown JF. Characterization of a calici-like virus (Newbury agent) found in association with astrovirus in bovine diarrhea. Infect Immun. 1984;43(1):133–8.
6. Cattoli G, De Battisti C, Toffan A, Salviato A, Lavazza A, Cerioli M, et al. Co-circulation of distinct genetic lineages of astroviruses in turkeys and guinea fowl. Arch Virol. 2007; 152(3):595–602.
7. Chu DKW, Poon LLM, Chan KH, Chen H, Guan Y, Yuen KY, et al. Coronaviruses in bent-winged bats (*Miniopterus* spp.). J Gen Virol. 2006;87(Pt 9):2461–6.
8. Chu DKW, Poon LLM, Guan Y, Peiris JSM. Novel astroviruses in insectivorous bats. J Virol. 2008;82(18):9107–14.
9. Chu DKW, Chin AWH, Smith GJ, Chan K-H, Guan Y, Peiris JSM, et al. Detection of novel astroviruses in urban brown rats and previously known astroviruses in humans. J Gen Virol. 2010;91(Pt 10):2457–62.
10. Englund L, Chriel M, Dietz HH, Hedlund KO. Astrovirus epidemiologically linked to pre-weaning diarrhoea in mink. Vet Microbiol. 2002;85(1):1–11.
11. Finkbeiner SR, Li Y, Ruone S, Conrardy C, Gregoricus N, Toney D, et al. Identification of a novel astrovirus (astrovirus VA1) associated with an outbreak of acute gastroenteritis. J Virol. 2009;83(20):10836–9.
12. Gavier-Widen D, Brojer C, Dietz HH, Englund L, Hammer AS, Hedlund KO, et al. Investigations into shaking mink syndrome: an encephalomyelitis of unknown cause in farmed mink (Mustela vison) kits in Scandinavia. J Vet Diagn Invest. 2004;16(4):305–12.
13. Gray EW, Angus KW, Snodgrass DR. Ultrastructure of the small intestine in astrovirus-infected lambs. J Gen Virol. 1980;49(1):71–82.
14. Grellet A, De Battisti C, Feugier A, Pantile M, Gradjean D, Cattoli G. Prevalence and risk factors of astrovirus infection in puppies from French breeding kennels. Vet Microbiol. 2012; 157(1–2):214–9.
15. Harbour DA, Ashley CR, Williams PD, Gruffydd-Jones TJ. Natural and experimental astrovirus infection of cats. Vet Rec. 1987;120(23):555–7.
16. Henriksen P. En oversigt over syndromet "Fedtede hvalpe" hos mink (a review of "sticky kit" syndrome in mink). Dansk Vet Tidskr. 1987;70:580–3.
17. Herbst W, Krauss H. Electron microscopy in the diagnosis of enteritis in cats. Tijdschr Diergeneeskd. 1989;114(6):328–33.
18. Hoshino Y, Zimmer JF, Moise NS, Scott FW. Detection of astroviruses in feces of a cat with diarrhea. Brief report. Arch Virol. 1981;70(4):373–6.

19. Järplid B, Mejerland T. Villous atrophy in the small intestine of mink kits with diarrhea. Scientifur. 22:157–60. Cited in Englund L, Chriel M, Dietz HH, Hedlund KO. Astrovirus epidemiologically linked to pre-weaning diarrhoea in mink. Vet Microbiol. 2002;85(1):1–11.
20. Jonassen CM, Jonassen TO, Grinde B. A common RNA motif in the 3' end of the genomes of astroviruses, avian infectious bronchitis virus and an equine rhinovirus. J Gen Virol. 1998;79(Pt 4):715–8.
21. Jonassen CM, Jonassen TO, Saif YM, Snodgrass DR, Ushijima H, Shimizu M, et al. Comparison of capsid sequences from human and animal astroviruses. J Gen Virol. 2001;82(Pt 5):1061–7.
22. Jonassen CM, Jonassen T, Sveen TØ, Sveen TM, Grinde B. Complete genomic sequences of astroviruses from sheep and turkey: comparison with related viruses. Virus Res. 2003; 91(2):195–201.
23. Kapoor A, Li L, Victoria J, Oderinde B, Mason C, Pandey P, et al. Multiple novel astrovirus species in human stool. J Gen Virol. 2009;90(Pt 12):2965–72.
24. Krishna NK. Identification of structural domains involved in astrovirus capsid biology. Viral Immunol. 2005;18(1):17–26.
25. Lukashov VV, Goudsmit J. Evolutionary relationships among Astroviridae. J Gen Virol. 2002;83(Pt 6):1397–405.
26. Luo Z, Roi S, Dastor M, Gallice E, Laurin MA, L'homme Y. Multiple novel and prevalent astroviruses in pigs. Vet Microbiol. 2011;149(3–4):316–23.
27. Madeley CR, Cosgrove BP. Letter: viruses in infantile gastroenteritis. Lancet. 1975; 2(7925):124.
28. Marshall JA, Healey DS, Studdert MJ, Scott PC, Kennett ML, Ward BK, et al. Viruses and virus-like particles in the faeces of dogs with and without diarrhoea. Aust Vet J. 1984; 61(2):33–8.
29. Marshall JA, Kennett ML, Rodger SM, Studdert MJ, Thompson WL, Gust ID. Virus and virus-like particles in the faeces of cats with and without diarrhoea. Aust Vet J. 1987;64(4):100–5.
30. Martella V, Moschidou P, Lorusso E, Mari V, Camero M, Bellacicco A, et al. Detection and characterization of canine astroviruses. J Gen Virol. 2011;92(Pt 8):1880–7.
31. Martella V, Moschidou P, Pinto P, Catella C, Desario C, Larocca V, et al. Astroviruses in rabbits. Emerg Infect Dis. 2011;17(12):2287–93.
32. Mittelholzer C, Englund L, Hedlund KO, Dietz HH, Svensson L. Detection and sequence analysis of Danish and Swedish strains of mink astrovirus. J Clin Microbiol. 2003;41(11): 5192–4.
33. Mittelholzer C, Hedlund KO, Englund L, Dietz HH, Svensson L. Molecular characterization of a novel astrovirus associated with disease in mink. J Gen Virol. 2003;84(Pt 11):3087–94.
34. Pantin-Jackwood MJ, Strother KO, Mundt E, Zsak L, Day JM, Spackman E. Molecular characterization of avian astroviruses. Arch Virol. 2011;156(2):235–44.
35. Quan PL, Wagner TA, Briese T, Torgerson TR, Hornig M, Tashmukhamedova A, et al. Astrovirus encephalitis in boy with X-linked agammaglobulinemia. Emerg Infect Dis. 2010; 16(6):918–25.
36. Rice M, Wilks CR, Jones BR, Beck KE, Jones JM. Detection of astrovirus in the faeces of cats with diarrhoea. N Z Vet J. 1993;41(2):96–7.
37. Rivera R, Nollens HH, Venn-Watson S, Gulland FM, Wellehan Jr JF. Characterization of phylogenetically diverse astroviruses of marine mammals. J Gen Virol. 2010;91(Pt 1):166–73.
38. Smits SL, van Leeuwen M, Kuiken T, Hammer AS, Simon JH, Osterhaus AD. Identification and characterization of deer astroviruses. J Gen Virol. 2010;91(Pt 11):2719–22.
39. Schneider RR, Hunter DB. Mortality in mink kits from birth to weaning. Can Vet J. 1993;34(3):159–63.
40. Snodgrass DR, Gray EW. Detection and transmission of 30 nm virus particles (astroviruses) in faeces of lambs with diarrhoea. Arch Virol. 1977;55(4):287–91.
41. Snodgrass DR, Angus KW, Gray EW, Menzies JD, Paul G. Pathogenesis of diarrhoea caused by astrovirus infections in lambs. Arch Virol. 1979;60(3–4):217–26.
42. Svennekjaer NC. Minksyngdomme og aktuelle problemer i minkavlen (mink diseases and current problems in mink farming). Dansk Pelsdyravl. 1954;12:384–404.

43. Toffan A, Jonassen CM, De Battisti C, Schiavon E, Kofstad T, Capua I, et al. Genetic characterization of a new astrovirus detected in dogs suffering from diarrhoea. Vet Microbiol. 2009; 139(1–2):147–52.
44. Tse H, Chan WM, Tsoi HW, Fan RY, Lau CC, Lau SK, et al. Rediscovery and genomic characterization of bovine astroviruses. J Gen Virol. 2011;92(Pt 8):1888–98.
45. Tzipori S, Menzies JD, Gray EW. Detection of astrovirus in the faeces of red deer. Vet Rec. 1981;108(13):286.
46. Vieler E, Herbst W. Electron microscopic demonstration of viruses in feces of dogs with diarrhea. Tierarztl Prax. 1995;23(1):66–9.
47. Williams Jr FP. Astrovirus-like, coronavirus-like, and parvovirus-like particles detected in the diarrheal stools of beagle pups. Arch Virol. 1980;66(3):215–26.
48. Woode GN, Bridger JC. Isolation of small viruses resembling astroviruses and caliciviruses from acute enteritis of calves. J Med Microbiol. 1978;11(4):441–52.
49. Woode GN, Pohlenz JF, Gourley NE, Fagerland JA. Astrovirus and Breda virus infections of dome cell epithelium of bovine ileum. J Clin Microbiol. 1984;19(5):623–30.
50. Woode GN, Gourley NE, Pohlenz JF, Liebler EM, Mathews SL, Hutchinson MP. Serotypes of bovine astrovirus. J Clin Microbiol. 1985;22(4):668–70.
51. Xiao J, Li J, Hu G, Chen Z, Wu Y, Chen Y, et al. Isolation and phylogenetic characterization of bat astroviruses in southern China. Arch Virol. 2011;156(8):1415–23.
52. Zhu HC, Chu DKW, Liu W, Dong BQ, Zhang SY, Zhang JX, et al. Detection of diverse astroviruses from bats in China. J Gen Virol. 2009;90(Pt 4):883–7.
53. Zhu AL, Zhao W, Yin H, Shan TL, Zhu CX, Yang X, et al. Isolation and characterization of canine astrovirus in China. Arch Virol. 2011;156(9):1671–5.

Chapter 9
Avian Astroviruses

Mary Pantin-Jackwood, Daniel Todd, and Matthew D. Koci

Abstract Avian astroviruses comprise a diverse group of viruses affecting many avian species and causing enteritis, hepatitis, and nephritis. To date, six different astroviruses have been identified in avian species based on the species of origin and viral genome characteristics: two turkey-origin astroviruses [Turkey Astrovirus type 1 (TAstV-1) and type 2 (TAstV-2)]; two chicken-origin astroviruses [Avian Nephritis Virus (ANV) and Chicken Astrovirus (CAstV)]; and two duck-origin astrovirus [Duck Astrovirus type 1 (DAstV-1) and type 2 (DAstV-2)]. ANV has also been detected in turkeys, ducklings, pigeons, and guinea fowl; and TAstrovirus-2-like viruses have also been found in guinea fowl. Astroviruses are commonly associated with enteric disease syndromes in poultry including runting-stunting syndrome of broilers (RSS), poult enteritis complex or syndrome (PEC or PES), poult enteritis mortality syndrome (PEMS), and enteritis in guinea fowl. The molecular characterization of the different avian astroviruses shows great genetic variability among each type, and this variability influences the ability to detect these viruses by molecular and serological techniques. In this chapter, we review the different aspects related to avian astroviruses, including molecular biology, pathogenesis, diagnosis, and control.

M. Pantin-Jackwood, D.V.M., M.S., Ph.D. (✉)
Southeast Poultry Research Laboratory, United States Department of Agriculture,
Agricultural Research Service, 934 College Station Road, Athens, GA 30605, USA
e-mail: mary.pantin-jackwood@ars.usda.gov

D. Todd, B.Sc., Ph.D.
Virology Branch, Agri-Food and Biosciences Institute,
Stoney Road, Belfast, BT4 3SD, UK
e-mail: Daniel.Todd@afbini.gov.uk

M.D. Koci, Ph.D.
Department of Poultry Science, North Carolina State University,
Campus Box 7608, Raleigh, NC 27695, USA
e-mail: mdkoci@ncsu.edu

S. Schultz-Cherry (ed.), *Astrovirus Research: Essential Ideas, Everyday Impacts, Future Directions*, DOI 10.1007/978-1-4614-4735-1_9,

Introduction

Avian astroviruses comprise a diverse group of viruses affecting many avian species and causing enteritis, hepatitis, and nephritis. Reports of astroviruses as cause of illness in birds predate that of mammals. In 1965, a fatal hepatitis was described in ducklings [2] and was eventually shown to be caused by an astrovirus [22, 23]. Another astrovirus, avian nephritis virus (ANV), was first isolated in 1976 and was associated with growth depression and kidney lesions in young chickens. Initially classified as an *Enterovirus*, it was later reclassified as an astrovirus following the complete sequencing of the viral genome [30]. To date, six different astroviruses have been identified in avian species based on the species of origin and viral genome characteristics: two turkey-origin astroviruses [Turkey Astrovirus type 1 (TAstV-1) and type 2 (TAstV-2)]; two chicken-origin astroviruses [ANV and Chicken Astrovirus (CAstV)]; and two duck-origin astrovirus [Duck Astrovirus type1 (DAstV-1) and type 2 (DAstV-2)]. ANV has also been detected in turkeys [18, 70], ducklings [6], pigeons [45, 116], and guinea fowl [8]; and TAstV-2-like viruses also have been found in guinea fowl [9, 10].

As in mammals, astroviruses are commonly associated with enteritis in avian species. Enteric diseases cause substantial economic losses in commercial poultry, and many enteric disease syndromes have been described including runting-stunting syndrome of broilers (RSS), poult enteritis complex or syndrome (PEC or PES), and poult enteritis mortality syndrome (PEMS). Astroviruses are among the most common viruses found in cases of PEC, PES, and PEMS in turkeys [2, 3, 34–37, 42, 68, 69, 75, 76, 79, 81, 99]. Astroviruses have also been isolated from cases of RSS in broilers and have been previously associated with poor weight gain, enteric disease, and kidney disease in chickens [4, 15, 27, 83, 85, 86, 105], as well as in guinea fowl suffering from enteritis [10].

Astroviruses were found in turkeys in the UK in 1980 by McNulty et al. [56], and later in the 1980s, in the USA (TAstV-1) [75, 79, 80]. A second TAstV, antigenically and genetically distinct from the previously described isolates, was later characterized [44] and this and similar astroviruses have been designated TAstV-2 [70, 71, 98]. Beyond this classification, data from genetic analyses of field viruses detected in turkeys indicate the circulation of multiple lineages of TAstV-2 [9, 18, 33, 70, 94].

Avian nephritis virus has also been shown to be widely distributed in chicken flocks [30, 53, 70, 103]. Under field conditions, clinical signs associated with infection with this virus broiler chickens have varied from subclinical to outbreaks of RSS and baby chick nephropathy [26]. A second type of astrovirus affecting chickens was shown to be antigenically and genetically distinct from ANV and was named chicken astrovirus (CastV, 4). CAstVs that share high levels of nucleotide sequence identity with this first characterized CAstV have been detected in the USA in broiler chickens affected with RSS [67]. Recently, molecular and antigenic characterization of enterovirus-like viruses (ELVs) demonstrated that they were CAstV, some of them closely related to the first characterized CAstV, and some of them more distantly related [102].

The molecular characterization of the different avian astroviruses shows great genetic variability among each type, and this variability influences the ability to detect these viruses by molecular and serological techniques. It is known that ANV and TAstV-2 each comprise at least two serotypes but many more are suspected. Some studies have shown that TAstV-2, ANV, and CAstV display substantial levels of capsid protein sequence variability, possibly indicating that there are numerous serotypes of each virus [33, 71, 72, 88, 94, 104]. Concomitant infections with different strains have been reported for avian astroviruses and this can result in recombination between strains [71, 72, 94, 104]. Recombination is likely to provide an important mechanism by which the sequence diversity of avian astroviruses is increased. In this chapter, we will review the different aspects related to avian astroviruses, including molecular biology, pathogenesis, diagnosis, and control.

Virus Aspects

Genomic Organization of Avian Astroviruses

The complete genomic sequence of 22 avian astrovirus isolates have been reported, including the genomes of TAstV-1,TAstV-2, ANV, DAstV, and CAstV [20, 30, 38, 39, 44, 94, 113, 115]. Like all other members of the *Astroviridae* family, avian astroviruses or avastroviruses are composed of a positive sense RNA ranging from 6,623 to 7,722 nt in length, excluding the poly-A tail (Fig. 9.1). While the specific lengths of individual regions vary among isolates, in general each comprises a short 5′ untranslated region (UTR), a large open reading frame (ORF) predicted to encode several nonstructural proteins (ORF1a), a second reading frame predicted to encode the RNA dependent RNA polymerase (ORF1b), a third reading frame which encodes the capsid protein (ORF2), and a 3′UTR, followed by a poly-A tail (Fig. 9.1).

Interestingly, while these basic features are well conserved across all of *Astroviridae*, examination of each of the reported genomes demonstrates essentially three basic genome configurations within the avian astroviruses. The ANV-like genome in which ORF1a and ORF2 are in the same frame and ORF1b is in the +1 reading frame; the TAstV-like genome in which ORF1b and ORF2 are both in the +1 reading frame; and the DAstV-like genome in which each ORF is in a distinct reading frame (Fig. 9.1). In addition to these organizational differences, the TAstV-2-like viruses are the only members of Avastrovirus missing the conserved s2m RNA stem-loop motif [44, 94]. While not unprecedented, as the bat, rat, and MBL1 astroviruses are missing this motif [11, 12], its absence suggests this feature is not essential. The prevalence of the s2m motif among *Astroviridae* as well as other viruses led Monceyron et al. to propose that it was introduced into an astrovirus progenitor via a recombination event between non-related viruses [62]. The biological significance of the loss of s2m or the differences in the organization of the ORFs is unclear, but may reflect differences in evolutionary pressures.

Phylogenetic comparison of the Avastrovirus genomes demonstrates that individual isolates cluster based primarily on the host they have been described to infect. One exception is that of TAstV-1, which is found more closely associated with

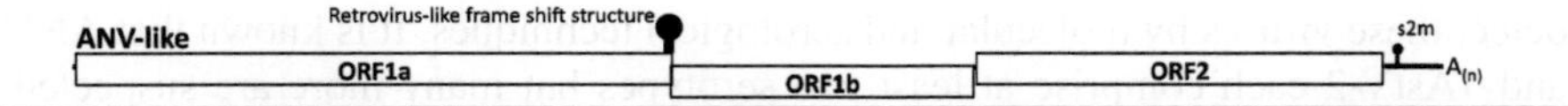

Isolate	Host	5′UTR	ORF1a	ORF1b	ORF2	s2m	3′UTR	Reference
ANV G-4260	Chicken	1-13	14-3031	3022-4551	4571-6622	6706-6740	6623-6927	6
Sichuan54	Chicken	1-13	14-3055	3046-4575	4595-6637	6720-6754	6638-6941	20
Pigeon ANV	Pigeon	1-13	14-3031	3022-4551	4571-6613	6696-6730	6614-6916	21

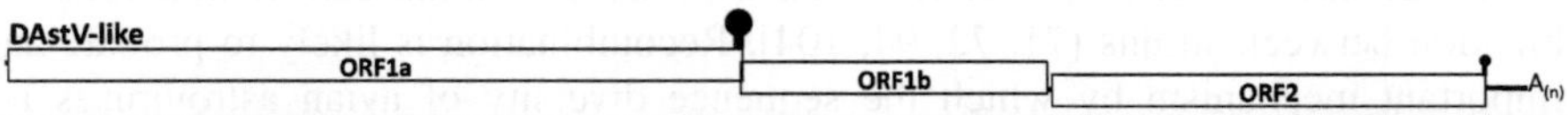

Isolate	Host	5′UTR	ORF1a	ORF1b	ORF2	s2m	3′UTR	Reference
C-NGB	Duck	1-22	23-3745	3736-5286	5310-7505	7486-7520	7506-7722	4

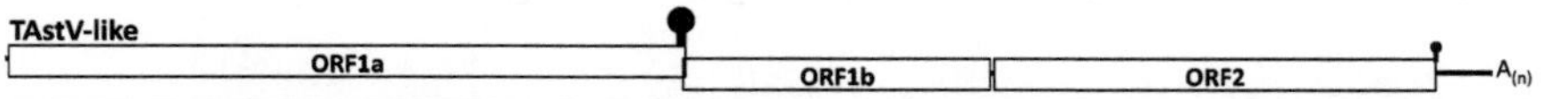

Isolate	Host	5′UTR	ORF1a	ORF1b	ORF2	s2m	3′UTR	Reference
TAstV-1/OH	Turkey	1-10	11-3310	3301-4839	4858-6873	6875-6909	6874-7003	8
CAstV/GA2011	Chicken	1-21	22-3441	3432-4982	5007-7238	7216-7253	7239-7520	9
TAstV-2/NC/99	Turkey	1-20	21-3398	3388-4936	4955-7129	NA	7130-7325	10
TAstV-2/MN/01	Turkey	1-20	21-3401	3391-4939	4967-7144	NA	7145-7337	18
TAstV-2/AR/98	Turkey	1-20	21-3398	3388-4936	4964-7141	NA	7142-7333	18
TAstV-2/TX/00	Turkey	1-20	21-3401	3391-4939	4967-7138	NA	7139-7331	18
TAstV-2/CO/01	Turkey	1-20	21-3398	3388-4936	4955-7135	NA	7136-7328	18
TAstV-2/VA/99	Turkey	1-20	21-3386	3376-4924	4952-7117	NA	7118-7309	18

Fig. 9.1 Genomic organization and lengths of regions for each fully characterized avastrovirus genome. Diagrams represent the three basic organizations of the avian astrovirus genomes, and list the specific length of each feature of the fully characterized avian astrovirus isolates. The numbers presented for ORF1a and ORF2 are the nucleotide positions of the start and stop codons. The numbers presented for ORF1a are the nucleotides from the beginning of the "slippery sequence" (AAAAAAC) to the ORF1a stop codon

chicken astroviruses than the other turkey astroviruses (Fig. 9.2). This discrepancy, however, is not unexpected as TAstV-1 has been classified as a member genogroup I (see Taxonomy) with the ANV-like viruses based on phylogenetic comparisons of the capsid genes [6]. The one non-ANV-like CAstV isolate included in this alignment, however, was found to group between genogroups I and II. At present, it is unclear if this represents a distinct genogroup or an evolutionary transition as it has some elements more consistent with genogroup I member TAstV-1 (genomic organization, Fig. 9.1) and others more consistent with genogroup II members TAstV-2 and DAstV (overall length and 5′ UTR, Figs. 9.1 and 9.3).

RNA Functional Elements

Comparison of the 5′ UTR of the different avastrovirus isolates demonstrates a conserved motif in spite of the differences in the lengths between the different viruses (Figs. 9.1 and 9.3). The conserved motif is even more prominent when viewed in

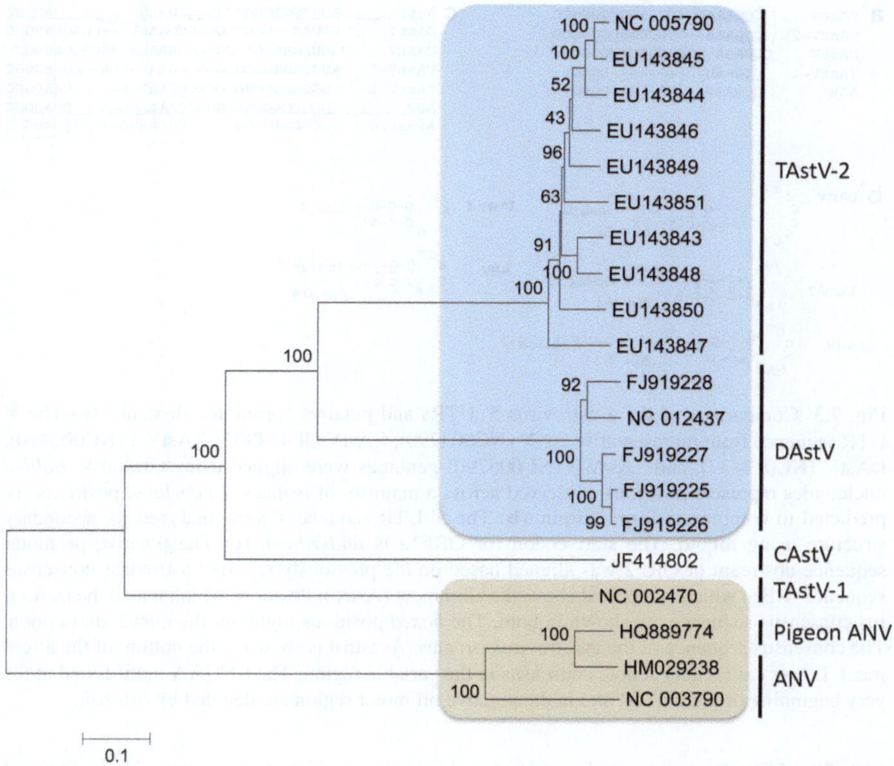

Fig. 9.2 Phylogenetic relationship of avastroviruses. The full-length genomes of each of the 22 avastroviruses were aligned using ClustalW2, and the phylogeny estimated neighbor using the neighbor joining method and 1,000 bootstrap replicates were calculated. The viruses that have been assigned to one of the two genogroups have been highlighted (genogroup I = *brown*, genogroup II = *blue*)

the context of its predicted secondary structure (Fig. 9.3b). The conserved stem-loop structure has been proposed to function as a promoter element for RNA transcription [38, 107]. This idea is supported by the fact that a similar highly conserved sequence can be found immediately upstream of ORF2 in the area where RNA transcription of the subgenomic message would begin (Fig. 9.3c). The function of these elements has not been demonstrated experimentally in any astrovirus isolate; however, given their conservation within the genome as well as across *Astroviridae* this suggests a functional role.

The formation of another RNA structural element, the retrovirus-like frame shift stem-loop, has been a defining characteristic of *Astroviridae* [32]. The presence or absence of this structure, along with the other basic genomic organization, has been one of the items used in the classification of novel viruses to the family [20, 30, 39, 44]. The stem-loop formed has been described to play a key role in the translation of the ORF1b as part of the ORF1a polyprotein in human astrovirus infected cells

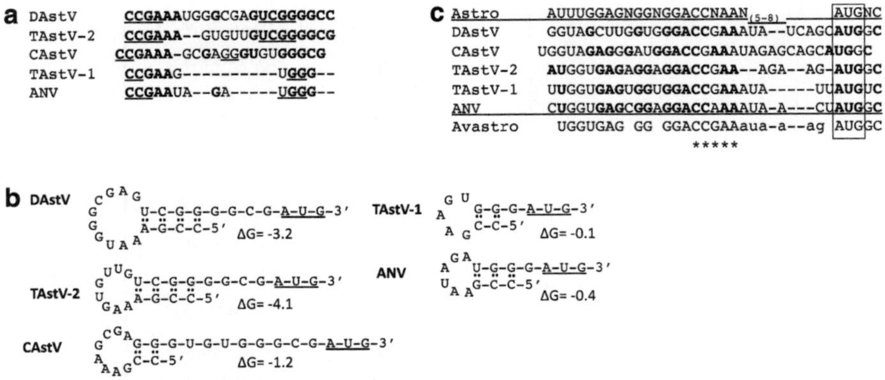

Fig. 9.3 Comparison of the avastrovirus 5′ UTRs and putative "promoter element." (**a**) The 5′ UTR sequence from representative ANV (NC003790), CAstV (JF414802), TAstV-1 (NC002470), DAstV (NC012437), and TAstV-2 (NC005790) genomes were aligned using ClustalW. *Bolded* nucleotides represent positions conserved across a majority of isolates. *Underlined* positions are predicted to complex to form hairpin. (**b**) The 5′ UTR sequences were analyzed for secondary structure using mFold. The start codon for ORF1a is *underlined*. (**c**) The putative promoter sequence upstream of ORF2 was aligned based on the previously reported astrovirus consensus sequence (REF) which is reported above the alignment (Astro). Positions which match the astrovirus consensus sequence are shown in bold. The *boxed* positions highlight the ORF2 start codon. The consensus sequence of the avastrovirus isolates (Avastro) is shown at the bottom of the alignment. Lower case letters denote consensus in the variable region. The CCGAA motif found at the very beginning of the 5′ UTR and in the putative promoter region are denoted by *asterisk*

[48–50, 55]. Experimental evidence demonstrated that the stem-loop formed allowed for a frame shift to occur upstream of the ORF1a stop-codon, and that brought ORF1b into frame with ORF1a resulting in the translation of one polyprotein [21, 49]. Based on this observation, it has been assumed that the presence of this structure in Avastrovirus genomes meant a similar mechanism allowed for the translation of the avian astrovirus RNA-dependent RNA polymerase (RdRp). Recently, studies by Kang et al. have suggested that the avian astroviruses may not use this strategy [39]. Comparison of the stem-loop predicted for human astrovirus to that of the avian viruses demonstrates differences in their ability to induce the needed frame shift (Fig. 9.4). In human astrovirus, the loop structure is between the "slippery sequence" (AAAAAAC) and the ORF1a stop codon (Fig. 9.4). Conversely, in the avian astroviruses, the ORF1a stop codon is immediately downstream of the "slippery sequence," and upstream of the stem loop structure (Fig. 9.4). Therefore, any frame shift induced may not allow for translation of ORF1a to read through to ORF1b. Kang et al. proposed that translation of ORF1b in the CAstV could initiate via a start codon just upstream of the "slippery sequence" [39]; however, this is the only avian virus to have a start site immediately upstream of the "slippery sequence" and would require the stem-loop structure not to be formed. Examination of these viruses for other alternative start sites revealed the first ATG in ORF1b was between 68 and 257 nt downstream from the ORF1a stop codon (Fig. 9.4). It is possible that a more complex stem-loop and frame shift structure is produced in these viruses that

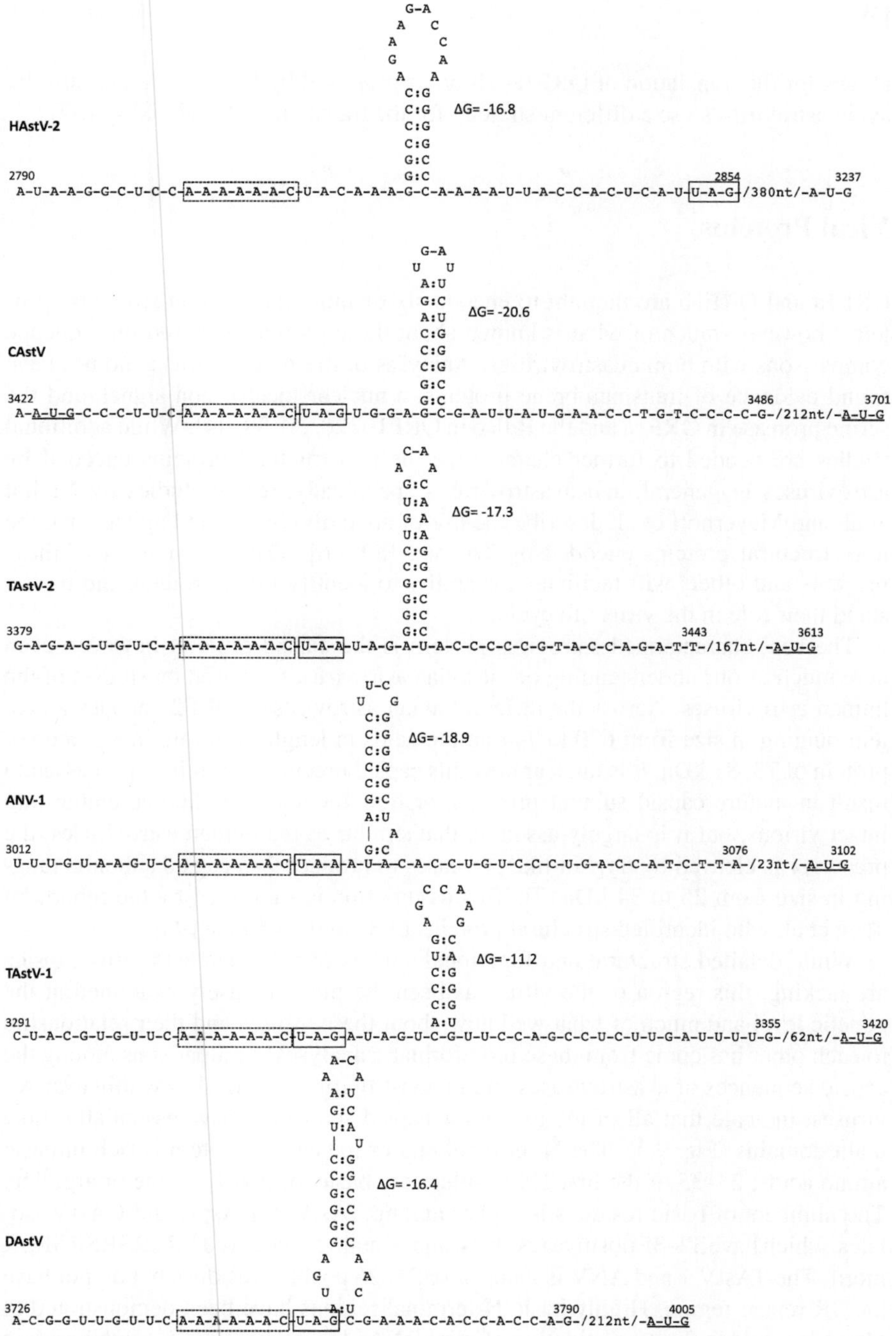

Fig. 9.4 Avastroviruses have conserved frame shift structure, but may have alternative translational strategy. The ORF1a-ORF1b region reported to form a frame shift structure for the translation of ORF1b was analyzed by mFold for ANV (NC003790), ChAstV (JF414802), TAstV-1 (NC002470), DAstV (NC012437), and TAstV-2 (NC005790) genomes, and compared with HAstV-1 (AY720892). For each genome, the predicted structure formed from a 65 nt region starting 10 nt upstream of the slippery sequence (AAAAAAC, highlighted by *dashed box*). The ORF1a stop codon is highlighted by a *solid box*, and potential start codons in the ORF1b frame are *underlined*

allows for the translation of ORF1a–1b or, as proposed by Kang et al., perhaps the avian astroviruses use a different strategy for the translation of their RdRp [39].

Viral Proteins

ORF1a and ORF1b are thought to encode six or more mature non-structural proteins; however, much of what is known about these proteins is based on sequence comparisons with human astroviruses. Analyses of the Avastrovirus genomes have found evidence of transmembrane proteins, a nuclear localization signal, and the serine protease in ORF1a and the RdRp in ORF1b [20, 30, 38, 44]. While additional studies are needed to further characterize the nonstructural proteins encoded by astroviruses in general, and avastroviruses specifically, recent studies by Nighot et al. and Meyerhoff et al. describe the use of an antibody directed against one the non-structural proteins encoded by TAstV-2 [59, 66]. The development of these reagents and others will facilitate our ability to identify these proteins and understand their role in the virus life cycle.

The best characterized astrovirus protein is the capsid protein; however, even here much of our understanding of the avian astroviruses is based on studies of the human astroviruses. Across the different avian astroviruses, ORF2 encodes a protein ranging in size from 670 to 744 amino acids in length resulting in a precursor protein of 73–81 kDa. It is unclear how this capsid precursor protein is processed to result in mature capsid subunit proteins, or how the subunits then assemble into intact virions, but it is largely assumed that similar to the human astroviruses, the precursor is cleaved by trypsin-like proteases to produce 2–3 capsid subunits ranging in size from 25 to 34 kDa [7]. This assumption is supported by the reports of Tang et al. who identified structural proteins of 31 and 34.5 kDa [98].

While detailed structural and functional studies of the avian astrovirus capsids are lacking, this region of the virus has been the most intensely examined at the genetic level and much of what we know about these viruses and their relationship to each other has come from these bioinformatic analyses. Comparisons among the capsid sequences of avastroviruses and mamastroviruses, as well as within avastroviruses, indicate that all of the astrovirus capsid sequences have essentially three main domains (Fig. 9.5). The N-terminal end of the capsid protein is rich in basic amino acids; 23–35 of the first 100 residues are basic, primarily lysine or arginine. The numbers of basic residues is higher among TAstV-2, DAstV, and CAstV isolates, which have 32–35 positive residues, and contain a conserved SRSRSRSRSRSR motif. The TAstV-1 and ANV isolates have 23–25 positive residues but do not have the SR repeat region. Highly basic N-terminal regions have been demonstrated to play a key role in the encapsidation of viral RNAs for several viruses, and therefore it is assumed that these residues play a similar role in astroviruses [46].

Analysis of the rest of the avian astrovirus capsid protein demonstrates, as previously described for human astroviruses, the N-terminal half of the protein is far

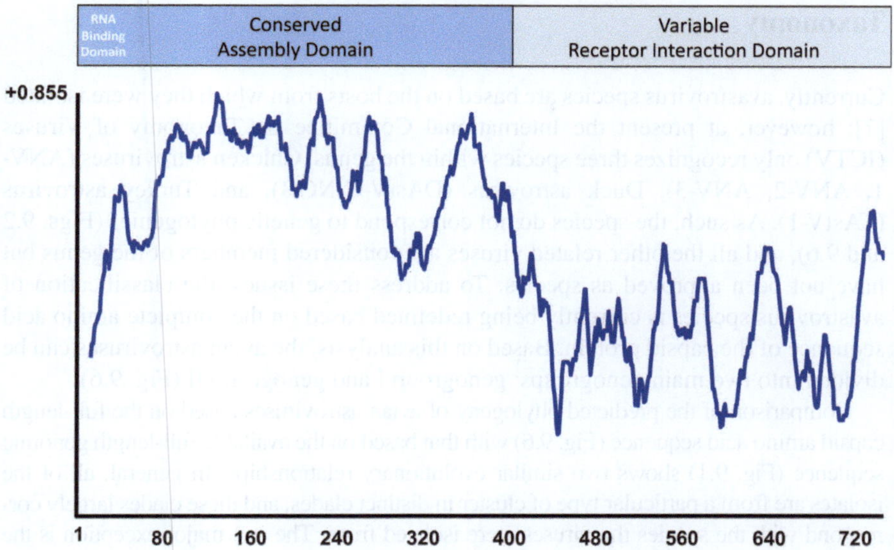

Fig. 9.5 Avastrovirus capsid domains. The relative position of the three functional domains of the astrovirus capsid protein, as described by Krishna (2005), are represented by the *blue bar* and is superimposed on percent similarity plot demonstrating the level of amino acid homology across 34 avastrovirus capsid sequences (see Fig. 9.5). The similarity plot was constructed using Vector NTI 10.0.1 and using a 25 amino acid window

more conserved than the C-terminal half (Fig. 9.5). Analysis of the human astroviruses indicate that the more conserved region of the capsid is important in virion assembly while the more variable region is found on the outer surface of the virion and is involved in interacting with cellular receptors and the immune system [46]. This suggests that these different regions of the capsid protein are under different evolutionary pressures. The first approximately 400 residues are important for encapsidation of the viral RNA and formation of the virion. As such, there are strong structural constraints to the sequence and therefore potential selection against amino acid substitutions. Conversely, the last approximately 300 residues would be exposed to antibodies produced by the host as well as binding to potential new host cells. Especially in the context of interspecies transmission, which appears to occur with some regularity among the avian astroviruses (see "Epidemiology" and "Pathogenesis" sections), one could imagine considerable evolutionary pressure driving amino acid changes in this region. Studies by Strain et al. support this idea [94]. Strain et al. identified 12 codons within the TAstV-2 capsid with evidence of positive selection. Among these 12 codons, 10 were located in the C-terminal half. Additional studies are required to better understand the roles each of these regions play in the life cycle of avian astroviruses and specifically how these different roles affect their evolution.

Taxonomy

Currently, avastrovirus species are based on the hosts from which they were isolated [7]; however, at present the International Committee on Taxonomy of Viruses (ICTV) only recognizes three species within the genus: Chicken astroviruses (ANV-1, ANV-2, ANV-3), Duck astrovirus (DAstV-C-NGB), and Turkey astrovirus (TAstV-1). As such, the species do not correspond to genetic phylogenies (Figs. 9.2 and 9.6), and all the other related viruses are considered members of the genus but have not been approved as species. To address these issues, the classification of avastrovirus species is currently being redefined based on the complete amino acid sequence of the capsid protein. Based on this analysis, the avian astroviruses can be divided into two main genogroups: genogroup I and genogroup II (Fig. 9.6).

Comparison of the predicted phylogeny of avian astroviruses based on the full-length capsid amino acid sequence (Fig. 9.6) with that based on the available full-length genomic sequence (Fig. 9.1) shows two similar evolutionary relationships. In general, all of the isolates are from a particular type of cluster in distinct clades, and these clades largely correspond with the species the viruses were isolated from. The one major exception is the TAstV-1 isolate, which is found more similar to the ANV viruses (Figs. 9.2 and 9.6). However, more careful inspection of the branch order of the TAstV-2 isolates between Figs. 9.2 and 9.6 demonstrates potentially important differences. Most notably, in Fig. 9.6 the CO/2001 and MI/2001 isolates are on the same branch; however, in Fig. 9.2 they were found to be more divergent. This discrepancy is due to a potential recombination event which occurred between these two viruses at the beginning of ORF2 [94]. This results in two isolates which appear to be identical based on the capsid analysis, but, in fact, are more divergent, and suggests analysis of the capsid region in most cases may be sufficient to assign an individual isolate to a particular group; understanding how that isolate may be related to others within that group requires more extensive comparisons.

As we identify new types of avian astroviruses, such as the CAstVs, it may be necessary to adopt new criteria for determining which genogroup it should be assigned to. Based on the full-genome phylogenetic analysis (Fig. 9.2) CAstVs appear to be closer to the genogroup II (DAstV and TAstV-2) viruses than the genogroup I viruses (ANV and TAstV-1). This observation is supported by the alignment of the capsid sequences (Fig. 9.6); however, there appears to be considerably more capsid diversity among the CAstV isolates than other avian astroviruses. In fact, based on the capsid there appears to be two distinct groups (Fig. 9.6). Comparisons of these groups based on limited regions of ORF1b demonstrate far more homology among these viruses [90], providing more evidence that understanding how individual isolates are related to other astroviruses may require comparisons at the whole genome level.

Epidemiology

Information about the range of avian species infected with astroviruses and the prevalence of avian astrovirus infections has been accumulated from a relatively small number of serological studies and an increasing number of recent studies

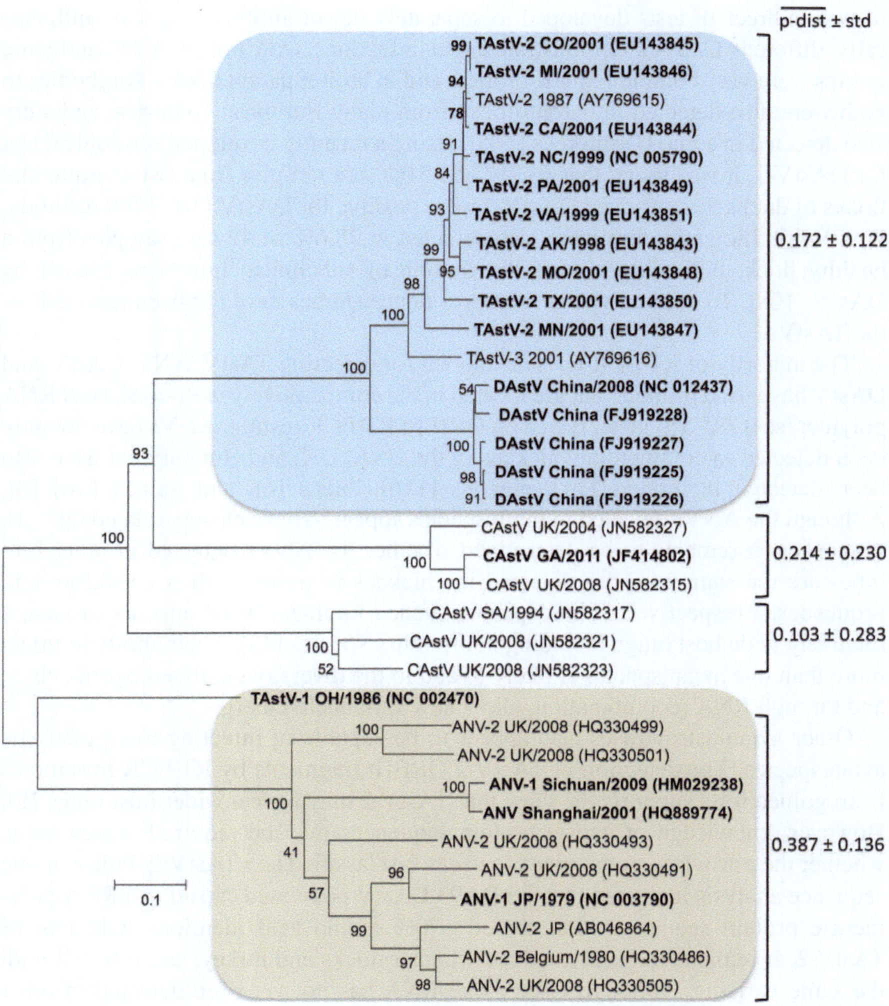

Fig. 9.6 Phylogenetic relationship of avastroviruses. The full-length amino acid sequences of the capsid proteins from 34 avastroviruses were aligned using ClustalW2, and the phylogeny estimated using the neighbor joining method and 1,000 bootstrap replicates were calculated using MEGA5. The viruses that have been assigned to one of the two genogroups have been highlighted (genogroup I = *brown*, genogroup II = *blue*). Isolates in *bold* denote those with full genomic sequence represented in Fig. 9.2. The mean amino acid genetic distance (p-dist) ± standard deviation for each of the major clades was calculated using MEGA5

based on the use of virus-specific RT-PCR tests. Using indirect immunofluorescence (IF) performed with the reference ANV-1 virus (G-4260), investigations in the 1980s and 1990s have shown that ANV infections are highly prevalent in commercial chickens in Japan and in several European countries [13, 31]. Antibody to ANV was also detected in turkey flocks and in SPF flocks [13, 58]. More recently, application of two indirect IF tests developed to separately detect antibodies to two antigenically different CAstVs demonstrated that infections with both CAstV antigenic groups were very common in UK broilers and in broiler parent flocks. Antibodies to both were also detected in parent flocks from many European countries, and were also detected in some UK turkeys [105]. Using a recently developed serological test for DAstV-1, it was found that 63.2% and 51% sera samples from two commercial flocks of ducks that survived DAstV-1 were positive for DAstV [108]. Interestingly, DAstV-specific antibodies were also detected in 28.6% of 42 sera samples from a healthy flock, indicating a wide distribution of subclinical infections caused by DAstV [108]. To our knowledge, seroprevalence studies have not been reported for the TAstVs.

The majority of RT-PCR tests developed for detecting TAstV, ANV, CAstV, and DAstV have used primers that are located in the comparatively well-conserved RNA polymerase (ORF 1b) gene. Based on ORF 1b RT-PCR testing, ANVs have not only been detected in commercial chickens in the USA, UK, and Hungary but have also been detected in turkeys [14], pigeons [116], ducks [6], and guinea fowl [8]. Although the ANVs detected in these species appear to have closely related ORF 1b sequences, it remains to be determined whether the ANVs detected in other bird types are the same as those detected in chickens in terms of their capsid protein sequences. Irrespective of the capsid sequence findings, ANV appears to have a relatively wide host range including non-poultry species [45]. Their ability to infect more than one avian species is likely to add to the diversity exhibited by this virus, and through RNA recombination, allow new variants to emerge.

Other avian astroviruses also appear to be capable of infecting more than one avian species. The detection of TAstV-2 ORF1b fragments by RT-PCR in samples from guinea fowl supports the view that TAstV-2 may have a wider host range [9]. However, knowledge of capsid protein sequences may be required to determine whether the astrovirus from guinea fowl can be classified as a TAstV-2. Full-genome sequence analysis indicated that the DA93 DAstV possessed capsid and RNA polymerase protein sequences that shared ~70% amino acid identities with that of TAstV-2. It remains to be determined whether ducks and turkeys are infected with the same variants. Although CAstV-like RNA has not yet been detected in other avian species, the detection by indirect IF of CAstV antibodies in turkeys suggests these birds could be infected with CAstV or antigenically related viruses [105]. The host range of TAstV-2 and CAstV appears to extend beyond turkeys and chickens, respectively, and, like ANV, may further extend to non-poultry species.

Although infections with ANV and CAstV are highly prevalent in commercially reared chickens, and infections with TAstVs are also highly prevalent in commercially reared turkeys, there is little or no information about the prevalence of these viruses in other avian species or about the prevalence of DAstVs in ducks. RT-PCR detection rates of 96% and 100% were obtained for CAstV and ANV when sample sets ($n = \sim 50$) from UK broiler flocks affected by enteritis and growth retardation were investigated [88, 103]. Although large numbers of samples from "normal" chickens have not yet been tested, RT-PCR findings to date suggest ANV and CAstV infections are endemic in chickens. TAstVs have a wide geographic distribution and are considered one of the most prevalent virus infections in 1–5-week-old poults suffering from diarrhea [34, 67, 70, 76, 77, 79, 94]. The prevalence of TAstV varies from 40% to 80% in symptomatic poults [9, 18, 34, 76] to less than 50% in healthy flocks [34, 76]. Field observations indicate that TAstV-2 is currently more prevalent in turkeys than TAstV-1 [9, 70] and its occurrence is often associated with other enteric viruses [34, 67, 69, 76].

The application of RT-PCR tests to longitudinal survey samples has shown that ANV and CAstV can be detected in the intestinal contents from chickens at all ages from 4 days of age to slaughter [89, 90, 103]. Given that as high as 10^8 virus RNA copies/extracted sampled are detected at some time points, it is likely that high levels of virus will be excreted, and, since astroviruses are relatively resistant to chemical and physical inactivation, environmental contamination will be high [89]. Another longitudinal survey conducted to detect enteric viruses in commercial turkey operations demonstrated the presence of TAstV-2, ANV, and TAstV-1 from 2 to 12 weeks of age [69].

With astroviruses being primarily considered as causing enteric infections, spread via the fecal–oral route is likely to be the major means of horizontal transmission. However, horizontal transmission by respiratory or reproductive (male to female via semen) routes cannot be ruled out. Young broiler or turkey poult flocks can become horizontally infected if the housing and equipment within is contaminated with virus due to inadequate cleaning and disinfection between flocks or if the virus is introduced into the house due to poor on-farm biosecurity. Young flocks can also become infected with virus that is acquired vertically. The detection of ANV infections in chickens maintained under quarantine conditions provided indirect evidence of vertical transmission [13], but recently, the detections of low levels of ANV in late-dead duck embryos [6] and chick embryos (Todd et al., unpublished) have provided direct evidence. CAstV has also been detected by RT-PCR in similar dead-in-shell embryo samples (Todd et al., unpublished). Astrovirus excreted from vertically infected chicks will spread horizontally within the flocks.

Pathogenesis

Diseases Caused by Avian Astroviruses

Avian astroviruses, as in mammalian species, have been not only associated with enteritis, but can also cause extra-intestinal diseases including hepatitis and nephritis. Direct correlation of astroviruses identified in feces with enteric disease in birds is complicated by the presence of other enteric pathogens. When separated from other viruses, experimental infection with avian astroviruses can cause illness but usually attenuated when compared to what is seen clinically. The difficulty in obtaining pure astrovirus isolates has hindered pathogenesis studies on some of these viruses.

Duck Astrovirus

Duck astrovirus (DAstV) is an important duck pathogen and causes a highly contagious and fatal hepatitis in ducklings [109]. To date, two different astroviruses causing hepatitis have been recognized in ducks, both of which were initially classified as picornaviruses and named duck hepatitis viruses 2 and 3 (DHV-2 and DHV-3). Duck hepatitis virus 2 (DHV-2), isolated originally by Asplin (1965) [1], was morphologically characterized as an astrovirus in 1984 and renamed duck astrovirus 1 (DAstV-1) [22, 23]. The virus was later definitively identified as an astrovirus [102] and completely sequenced [20]. Duck hepatitis virus 3 (DHV-3), first reported by Toth in 1969 [106] and characterized by Haider and Calnek in 1979 [24], was recently identified by sequencing as a second astrovirus in ducks (DAstV-2) [102].

DAstV-1 (DHV-2) can cause a severe acute disease in young ducks which can result in up to 50% mortality [109]. Infection occurs through oral and cloacal routes. Deaths occur within 1–4 days, usually within 1–2 h after the appearance of clinical signs, which include polydypsia with loose droppings, excessive urate excretion, and sometimes convulsions and acute opisthotonos. Affected ducks usually die in good condition and both the mean death time and the mortality rate (10–50%) depend on the age of the ducks [22]. Survivors excrete virus for at least 1 week after infection and rear normally, with little evidence of retarded growth. The target organs for DAstV-1 appear to be liver and kidneys. The liver is usually pale pink with hemorrhages, and the kidneys are often swollen. Microscopic changes are characterized by extensive necrosis of hepatocytes and bile duct hyperplasia [22]. DAstV-2 has a low pathogenicity for ducklings experimentally infected, and only ducklings appear to be infected by the virus. Mortality rarely exceeds 30% [109] and mature ducks are refractory to the disease [22].

In a recent study, very low mortality rates (4.67% and 0.63%) were observed in two flocks affected by DAstV-1 [108], in sharp contrast with the high mortality rate (up to 50%) in a recent Chinese outbreak and in the 1983 UK outbreak [20, 22]. This may suggest differences in pathogenicity among DAstVs. Co-infections of DAstV and other viruses may also play a role in high mortality observed in some outbreaks.

Turkey Astrovirus

Turkey astroviruses are commonly detected in commercial turkeys experiencing viral enteritis. TAstV-2 has been associated with many enteric disease syndromes in turkeys including poult enteritis complex or syndrome (PEC or PES), and poult enteritis mortality syndrome (PEMS) [8, 18, 34, 44, 70, 75, 112]. Clinical signs of the disease usually develop between 1 and 3 weeks of age and generally last up to 2 weeks [78]. Severity is usually mild to moderate and is characterized by diarrhea, listlessness, litter eating, and nervousness. Although mortality rate is low, morbidity occurring as decreased growth is of great economic concern [2, 78]. The exact role of TAstV-2 as the etiological agent of enteritis in turkeys is difficult to estimate since it can also be

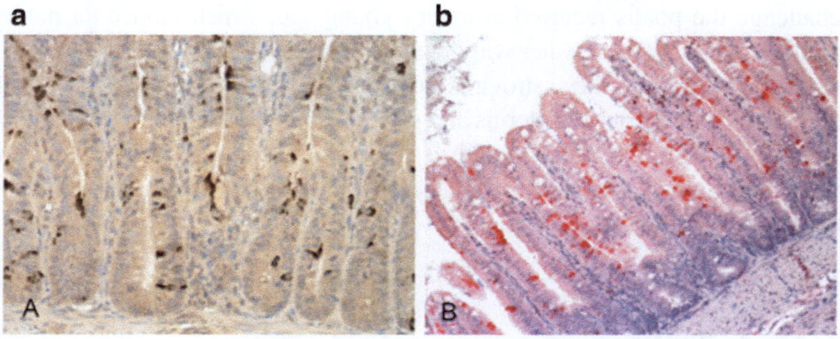

Fig. 9.7 Photomicrographs of jejunum sections from a turkey infected with TAstV-2, 3 days after inoculation; immunohistochemistry (IHC) and in situ hybridization (ISH) staining for detection of the virus. (**a**) Viral antigen staining present in the cytoplasm of the enterocytes at the middle section of the villi (*dark brown*); immunoperoxidase labeling, hematoxylin counterstain. (**b**) Viral RNA staining present in the cytoplasm of the enterocytes (*red*); digoxygenin labeling, BCIP-NBT staining. Magnification 200×. Pathogenesis of type 2 turkey astroviruses with variant capsid genes in 2-day-old specific pathogen free poults. M. Pantin-Jackwood, E. Spackman and J.M. Day. Avian Pathology 37, 193-201. Reprinted with permission of the publisher (Taylor & Francis Ltd, http://tandf.co.uk/journals)

detected in apparently healthy turkeys [69, 76]. TAstV-1 has also been detected in healthy turkeys and turkeys with enteric disease, but is less prevalent than TAstV-2.

Experimentally, young turkeys infected with TAstV-2 develop a diarrhea by 2 days post infection which continues for several days [41, 63, 68, 92]. TAstV-2 infection results in overall growth depression of infected birds. Common lesions observed at necropsy are dehydration, distended intestines filled with watery contents and undigested feed, and dilated ceca with foamy contents. Microscopic lesions present in the intestines consist of mild crypt hyperplasia, mild villous atrophy, and mild lymphocytic infiltration, and were most common in the jejunum [5, 41, 68]. Presence of the viruses has been demonstrated by immunohistochemistry (IHC) and by in situ hybridization (ISH) in both villi and crypt enterocytes in the duodenum, jejunum, ileum, and ceca (Fig. 9.7) [5, 41, 59, 66, 68]. Mild lymphocytic infiltration was also observed in other organs including pancreas, liver, spleen, and kidneys. Viremia occurred after infection with TAstV-2 [41, 68]; however, only intestinal cells appear to support viral replication. The possibility exists that low levels of viral replication may occur in other tissues and may not be detected with the tests used which would explain the extra-intestinal lesions observed.

Mild-to-moderate atrophy of the bursa of Fabricius and thymus was found in TAstV-2-infected poults [5, 41, 63, 68, 74, 81, 92, 99], and was associated with the depletion of lymphoid areas. The cause of these changes in the primary immune organs is unclear and could result in the immune suppression that some have reported with TAstV-2 infections (see section "Immunity"); however, the lymphoid depletion could also be a consequence of maldigestion and malabsorption leading to nutritional deficiencies in these rapidly growing poults.

Early mortality has been reported in poults experimentally inoculated with TAstV-2 viruses [68, 111]. The mortality in these cases most likely occurred because of the high viral challenge the poults received at a very young age, which induced a moderately severe lethargy and feed and water withdrawal that resulted in death. In an earlier study, the pathogenicity of two turkey astrovirus isolates was compared in older turkeys (14–28 days old). Both viruses caused enteritis and growth depression but no mortality [99].

Based on RT-PCR, TAstV-2-like viruses have also been detected in guinea fowl presenting enteric disease but these viruses require further genetic characterization to confirm their identity as TAstV-2 [10].

Avian Nephritis Virus

Avian nephritis virus (ANV) was first isolated from rectal contents of apparently healthy broiler chicks [110] and experimental infections revealed that it mainly results in a sub-clinical disease, although mild growth depression and mortality have been reported for infections with ANV strains including the serotype 1G-4260 strain (ANV-1) [29, 52, 110]. ANV typically causes histological changes in the kidneys although the viral antigen can be detected in other organs including liver, spleen, pancreas, jejunum, and rectum [27, 29, 84, 86]. Young chicks are the most susceptible and become resistant to disease after the first month of life [28]. ANV infections in chicks can vary from subclinical to outbreaks of runting stunting syndrome (RSS) and nephropathy [19, 31, 86, 95, 96, 110]. However, the role of ANV infection on RSS is not clear. Although ANV affected body gain in SPF chicks [85], an experimental study on the effect of ANV and several unclassified enterovirus-like viruses (ELVs) in commercial chicks demonstrated that only one ELV, later also characterized as an ANV, analyzed among the broad panel produced clinical signs in inoculated chicks [57]. Similarly, infection of 1-day-of-age chicks with ELVs, which were later identified as the FP3 and 612 isolates of CAstV and an ANV, showed that the ELVs are able to induce more severe lesions in the intestine, the kidneys, and the pancreas than the G-4260 strain of ANV [89]. Passive and active surveys indicate that ANV is widely distributed worldwide and that serological evidence can be found in turkeys as well [13, 16, 58, 65, 95]. Although an ANV genetically related to ANV-1 was detected in the feces of turkeys [70, 72], to date the infection has never been associated with the disease in turkeys. An attempt to infect poults with the chicken-derived ANV-1 strain was unsuccessful [61].

In chickens experimentally infected with ANV, the virus was first detected in feces 2 days after inoculation, with maximum virus shedding at 4–5 days after inoculation. The virus was widely distributed in tissues, with high titers in the kidney and jejunum and lower titers in the bursa of Fabricius, spleen, and liver [26]. Transient diarrhea and weight gain depression are common signs among 1-day-old infected chickens. At necropsy, mild-to-severe discoloration and swelling in the kidneys are observed, and in dead chicks visceral urate deposits are found [26]. The microscopic lesions described in infected chickens include necrosis and degeneration of epithelial cells of the proximal convoluted tubules of the kidney, interstitial lymphocyte infiltration, and moderate fibrosis. ANV virus particles and antigen were demonstrated in the degenerating kidney epithelium by electron microscopy and IF respectively [26].

Recently, novel ANVs have been isolated from chickens and turkeys suffering from runting and stunting-like and/or locomotory disease [15]. One of these isolates, isolate 19, induced diarrhea and runting and stunting in 3-week-old SPF layer chickens; however, the clinical tenosynovitis, which was seen regularly in the broilers and poults from the field, was not seen in the experimentally infected layers [15]. The isolation of these new avian astrovirus strains from chickens and turkeys with acute tenosynovitis and arthritis without the isolation of any other detectable pathogen suggests that they also might play a role in leg disorders. The pathogenicity of this virus was further demonstrated by the mortality and histopathological findings and seems to be broader than is reported for ANV-1 [15]. Many studies have shown that the pathogenicity of ANV-1 is limited to the kidney of chickens, in contrast to the damage observed in the kidney, pancreas, duodenum, thymus, bursa, and liver seen after infection with this new ANV [15]. Other ANVs have been shown to vary in pathogenicity and tropism. A Belgian PV2 isolate, originally recognized as an ELV, was shown to cause substantial growth retardation, and severe and long-lasting intestinal lesions following experimental infection [17]. Lesions were detected in the pancreas and proventriculus and were absent in the kidneys from chickens experimentally infected with PV2, suggesting that some ANV isolates such as PV2 may have different tissue tropisms [17].

The presence of ANV has also been demonstrated by RT-PCR in organ samples collected from 30-day-old ducklings and from dead-in-shell ducklings [6]. Histological examination of many tissues indicated that lesions were confined to the kidney and intestine. The finding of ANV in ducklings dead in shell indicates naturally occurring vertical infection and it is possible that ANV may play a role in the death of duckling embryos. ANVs were also detected by RT-PCR in 89% of fecal samples from young pigeons suffering from diarrhea, whereas samples from healthy pigeons were negative [115]. In another study, a very high prevalence of ANV was found among feral pigeons and wood pigeons [45]. Astrovirus was detected in almost all juvenile birds, while the detection rate among the adults was 62%. In this case, most birds appeared healthy, and it was not possible to correlate the presence of a virus to any illness in the birds. ANV-like viruses have also been detected in guinea fowl but the role they play in disease is not clear [8].

Chicken Astrovirus

The common detection of CAstV by RT-PCR in cases of broiler chickens affected by growth retardation suggests that infections with this avian astrovirus are contributing to poor growth performance [89]. High levels of CAstV RNA and CAstV antigen have also been detected in kidney tissue from chickens affected by nephritis and visceral gout (Todd, unpublished results).

Experimental infections of 1-day-of-age and 5-day-of-age SPF layer chickens with CAstV caused mild diarrhea and, except for some areas of limited damage at the base of the villi of the small intestine in some birds, no histological visible damage in the liver, kidney, spleen, thymus, or bursa was observed [4]. Used ELV-4 isolate 612, later typed as a CAstV [102], to infect commercial 1-day-old broilers with maternal antibodies against the virus and found only a growth depression of 11.3%, 20.4%, and 9.6% in the three trials that were performed [56]. Experimental infections of 1-day-old SPF layer chickens with CAstV strain 612 caused only very mild histological changes in the intestine and liver, and mild-to-moderate lesions in the kidney [87]. An experimental infection with the FP3 CAstV strain, which was originally known as ELV-3 and which is genetically and antigenically distinct from the 612 strain isolate, caused no clinical signs in 1-day-old broilers [57]. However, experimental infections of 1-day-old SPF layer chickens with the FP3 isolate caused mild-to-moderate histological changes in the kidney, pancreas, and intestine, with virus-specific antigen being detected in these tissues [87].

Mechanisms of Viral Pathogenesis

Our understanding of how astroviruses cause disease trails behind our knowledge of its molecular and epidemiologic properties and is confined mainly to work done with TAstV. Studies by Thouvenelle et al. and Reynolds et al. suggested that astrovirus infection induces changes in the expression of constituent enzymes of the brush border, specifically maltase, with a significant decrease in intestinal D-xylose absorption, resulting in disaccharide maldigestion, malabsorption, and subsequently osmotic diarrhea, even though it causes only mild microscopic lesions [75, 101]. The cellular mechanisms responsible for these changes are still unknown.

Most of the studies done with TAstV have demonstrated that astrovirus infection induces diarrhea in the absence of significant changes in the intestinal morphology. Furthermore, there is little evidence of inflammation and little or no increase in cell death following astrovirus infection [5, 41, 101], further suggesting astrovirus clinical signs result from submicroscopic changes in the intestine. Initial studies to identify physiological changes in infected poults demonstrated changes to ultrastructure of the epithelial cells including rearrangement of F-actin. These changes were associated with decreased apical expression of the sodium/hydrogen exchanger 3 (NHE3) with a concomitant decrease in sodium absorption [66]. Collectively, these data suggest that astrovirus infection induces a malabsorption, osmotic diarrhea, possibly through redistribution of specific brush border enzymes and transporters resulting from dysregulation of the actin cytoskeleton [64, 66, 75, 101].

Immunity

The immune response to astroviruses has not been well characterized in most species. In humans, it has been postulated that anti-astrovirus antibodies developed following natural infections are key mediators in protecting normal healthy adults

[47, 60]. In birds; however, this is far less clear and may differ among host species and/or types of avian astroviruses.

Studies with ANV, CAstV, and DAstV suggest that recovery from natural infection and/or vaccination leads to an immune response that protects the host from subsequent infection [1, 22, 28, 83, 109]. Multiple serotypes have been reported for each of these astrovirus types [4, 86, 104]. There are reports of at least two serotypes of TAstV-2; however, curiously there is no experimental evidence of an anamnestic response [41, 100]. This lack of an adaptive immune response has led several laboratories to investigate its effects on the immune system. These studies suggest TAstV-2 may induce immune dysregulation and/or localized inhibition [40, 41, 73, 74].

Studies by Qureshi et al. examined the response of lymphocytes from TAstV-2 or mock-infected poults to the mitogen concanavalin A [74]. These studies demonstrated a decrease in the *ex vivo* proliferative response to ConA by peripheral blood lymphocytes from TAstV-2 infected poults as compared with controls. Additionally, thymocytes and splenocytes that were exposed to the virus *in vitro* also demonstrated a reduced proliferative response [74]. In subsequent studies, Qureshi et al. reported TAstV-2 infections also resulted in impaired migration, phagocytic activity, and bacterial killing activity of innate immune cells, as compared with controls [73]. Functional examination of the innate immune cells (macrophages and heterophils) from TAstV-2 infected poults demonstrated reduced expression of the proinflammatory cytokines IL-1 and IL-6 [73]. Interestingly, studies by Koci et al. demonstrated significant increases in the levels of active TGF-β in the serum of TAstV-2 infected poults [41]. Collectively, these findings suggest that TAstV-2 infection leads to some sort of immune dysfunction, the specific mechanism of which is still not understood.

In spite of the immune inhibitory effects reported with TAstV-2 infection, the clinical disease is largely resolved within 12–14 days [41], suggesting that some anti-viral host factors are produced. Studies by Qureshi et al. and Koci et al. have both demonstrated that macrophages and adherent splenocytes from TAstV-2-infected poults produce more nitric oxide (NO) following *ex vivo* stimulation with lipopolysccharide than do cells from uninfected controls [40, 73]. NO has long been recognized as a potent anti-viral effector molecule; however, histological examination of TAstV-2 infected poults does not demonstrate large influxes of macrophages or other innate immune cells. To determine if NO could play a role in the regulation of TAstV-2 replication, the virus was exposed to NO donor compounds *in vitro* prior to inoculating embryos, and found to inactivate the virus [40]. Conversely, when embryos were pretreated with an inhibitor of the inducible nitric oxide synthase (iNOS) enzyme virus replication increased significantly [40, 51]. Finally, to determine if TAstV-2-infected intestines demonstrated an increase in the expression of iNOS, tissues from TAstV-2-infected turkey poults were stained for the presence of TAstV-2 antigen and turkey iNOS. These experiments demonstrated an increase in iNOS expression following TAstV-2 infection, with the majority of the staining localized to the intestinal epithelium suggesting that in the absence of a response from either innate or adaptive immune cells the intestine is capable of mounting its own innate defense [59].

Diagnosis

Astrovirus infections of commercially reared chickens and turkeys are endemic and this may also be the case with commercially reared ducks. Although infections are common, many are likely to be subclinical, with the incidence of clinical disease that can be directly attributed to a single avian astrovirus being relatively small. Therefore, with regard to diagnosis, it is important to differentiate whether astrovirus-related diseases are being diagnosed or whether astrovirus infections are being diagnosed.

The detection of clinical signs, gross pathological findings, and histological lesions is required before disease can be diagnosed. However, in the case of astroviruses, these in themselves may not be sufficient to identify the causal virus. For example, hepatitis in ducklings may be caused by one of the two different astroviruses: DAstV-1 and DAstV-2, originally known as duck hepatitis virus types 2 and 3, or the picornavirus, originally known as duck hepatitis virus type 1 [109]. Similarly, histological lesions consistent with nephritis in chickens can be caused by ANV, infectious bronchitis virus (IBV), and CAstV (Todd et al., unpublished results). In cases of hepatitis and nephritis, the presence of a specific astrovirus needs to be demonstrated in the affected tissue, preferably at the site of the histological lesion. This can be achieved using immunohistochemistry (IHC), undertaken with paraffin-embedded formalin-fixed tissue, or immunofluorescence (IF) performed with frozen tissue or tissue impression smears. However, due to the limited availability of avian astrovirus-specific antibodies, to date these methods have received limited application. IF is used to diagnose hepatitis in ducklings caused by DAstV-2 [24]. Very recently, a prototype immunoperoxidase-based IHC method has been applied to demonstrate the presence of CAstV antigen in nephritis-associated histological lesions (Todd et al., unpublished results; Fig. 9.8). This work suggested that CAstV alone was capable of causing nephritis in chickens and highlighted the need to identify the virus responsible in cases of nephritis and visceral gout. Imada et al. has reported that IF is a useful diagnostic procedure for ANV, detecting the group antigens and staining infected kidneys during the early acute phase of the disease [26]. *In situ* hybridization (ISH) can also be used to detect the presence of viral nucleic acid at the site of the histological lesion. Because probes generated from cloned virus-specific cDNA fragments may be more easily produced than virus-specific antibodies, ISH may prove to be an easier approach to develop. ISH has been used to demonstrate the presence of TAstV-2 RNA within the enterocytes present on the sides of the villi from virus-infected turkeys, but its use to detect other avian astroviruses has not yet been reported [5, 41, 68].

Diagnosing astrovirus infections is considerably easier than diagnosing astrovirus-associated disease. The method first used was negative contrast electron microscopy (EM), which depended on the visualization of the characteristic five- and six-pointed star morphology that characterizes astroviruses. EM performed with enteric samples has been used for the diagnosis of TAstVs and for the diagnosis of DAstV-1 in liver samples from ducks with hepatitis [23, 56]. However, astrovirus

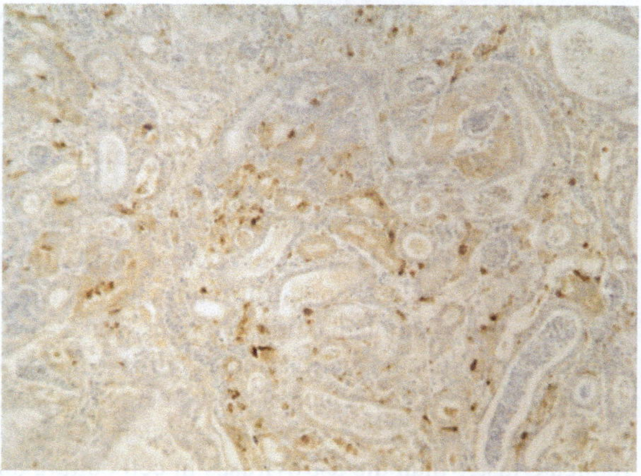

Fig. 9.8 Photomicrograph of a kidney section from a broiler chicken naturally infected with CAstV; immunohistochemistry (IHC) staining for detection of the virus. Viral antigen staining present in kidney epithelium (*dark brown*); immunoperoxidase labeling, hematoxylin counterstain. Magnification 200×

diagnosis by EM has a number of drawbacks, the most important being that some avian astroviruses fail to display the characteristic star morphology. This appears to be the case for ANV, DAstV-2, and CAstV, which are detected as small spherical viruses without obvious surface structure, and, on this basis, these were originally considered to be picornaviruses (ANV) or enterovirus-like viruses (CAstV, DAstV-2) [57]. Additional drawbacks of EM include its relative lack of sensitivity, and its unsuitability for the testing of large samples due to the time, labor, and equipment required.

Virus isolation in embryonated eggs or cell culture has been successful for most of the avian astroviruses including ANV, CAstV, TAstV, and DAstV. TAstV and DAstV can be propagated in turkey and duck embryos, respectively, and sometimes in chick embryos as well [98, 109]. Although the propagation of DAstV-1 (formerly DHV-2) in cell culture has been reported [4], the propagation of TAstV in cell cultures has not. The growth of high yields of TAstV isolates in turkey embryos has facilitated the production of purified virus, with which virus-specific antibodies have been prepared [97]. The astroviruses from chickens can be propagated in chicken embryos and in cell culture. CAstVs can be propagated in SPF chick embryos following yolk sac inoculation to produce variable effects including early embryo death, dwarfing, hemorrhages, edema, and feather changes [91]. ANVs can be propagated in SPF embryos following egg inoculation using a variety of routes. Isolation of ANV in cell cultures has been reported in a number of studies [19, 25, 110], and induces a cell rounding CPE in primary chicken kidney cells typically within 72 h post infection [26] . Experience from some laboratories suggests that some ANVs may be non-cultivable and the inconsistent growth of some ANVs in primary cell culture and embryos has prompted speculation that embryos containing ANV antibody may inhibit in vitro propagation (Todd et al. unpublished results). Baxendale and Mebatsion have reported the isolation and propagation of CAstVs in primary chick embryo liver cells and LMH cells, in which plaques similar to those produced by DAstV (DHV-2) were obtained [4]. The CAstV isolates have been used in virus neutralization tests, with which antibody to CAstV was detected [4]. There is also evidence that CAstVs can non-productively infect when inoculated into primary chick embryo liver cells [105]. In these cases, the inoculated CAstV undergoes partial replication, producing virus proteins that can be detected by indirect IF without producing infectious virus and CPE [105]. The effects produced by particular avian astroviruses in cell culture or in experimentally inoculated embryos are relatively nonspecific and cannot easily be distinguished from the effects caused by other avian astroviruses or other viruses. As such, the cultured virus needs to be identified; this can be accomplished through the detection of virus-specific antigens or nucleic acid. Although cultivable virus isolates are essential for biological characterization and pathogenesis studies, virus isolation is not routinely used to detect avian astroviruses, the principal reason being that astrovirus-containing samples are frequently contaminated with viruses such as adenoviruses or reoviruses, which replicate faster than the astroviruses in cell culture and/or in embryos. In addition, virus isolation and identification is time-consuming and labor-intensive, making it unsuitable for testing large samples.

A microtiter plate-based ELISA has been developed for the detection of TAstV-2 antigen in feces or intestinal content samples [97]. This test depended on the use of rabbit polyclonal antibodies that were raised by immunization with two different TAstV-2, which had been purified following replication in turkey embryos. Results compared favorably with those obtained by EM in terms of the numbers of positive samples recorded. Despite their convenience and low cost, the absence of virus-specific antibodies for other avian astroviruses has restricted the development of antigen-detecting ELISAs.

Attention has focused increasingly on the use of molecular diagnostic tests for detecting avian astroviruses. Since the first report describing an RT-PCR test for detecting TAstV [43], conventional and/or real-time tests have been described for ANV [14, 54, 103], CAstV [70]; [89], and DAstV [6] as well as TAstV-1 and TAstV-2 [14, 41, 93, 97]. The primers for these tests are usually located in the genomic regions encoding nonstructural proteins such as the RNA polymerase (ORF 1b), which are more conserved than the capsid protein encoding region (ORF 2). However, the high levels of sequence diversity displayed by avian astroviruses species make it difficult to select primers that will detect all variants within one type. Therefore, primer design should be based on comparing sequences from as large a number of virus variants as possible and the need to incorporate degenerate primers recognized. The importance of primer selection was highlighted when it was shown that the RT-PCR test using primers located in the 3′ untranslated region (3′ UTR) was more successful in detecting ANV-positive samples than tests based on primers located in the nonstructural protease or RNA polymerase [89, 103]. Similarly, CAstV from all genetic groupings was detected by the conventional and real-time RT-PCR tests described by Smyth et al. [88, 89], whereas one CAstV grouping was not detected by another published test [14]. Applications of these molecular tests have demonstrated that TAstV, ANV, and CAstV are highly prevalent in samples from their respective hosts, and there is a strong likelihood that the vast majority of commercial chicken and turkey flocks will become infected at some point in their lifetimes [18, 69, 70, 89]. Therefore, since all enteric samples are likely to be positive using these highly sensitive RT-PCR methods, knowing that an enteric sample is astrovirus positive is of questionable importance. However, with regard to the impact that the infection may be having on the flock or bird, there may be considerably more value in knowing the amounts of virus present, and in knowing whether internal tissues such as the kidney or liver are virus positive.

In comparison with other avian viruses, the serological diagnoses of avian astroviruses have received relatively little attention. A major reason for this is the limited number of avian astrovirus strains that have been adapted to cell culture. Although a number of ANV isolates can be propagated in cell culture, with very few exceptions cell culture isolations of CAstVs, TAstVs, and DAstVs have not been described. The lack of cell culture methods limits the development of indirect IF assays and virus neutralization tests to analyze chicken, turkey, and duck sera for antibodies. Additionally, the development of other serological assays such as ELISAs requires the ability to produce ample quantities of virus and virus antigen. With limited cell culture propagation methods, only a few laboratories have developed these assays for understanding astrovirus serology. An ELISA for detecting ANV antibody was described by Decaesstecker and Meulemans [16]. Sellers et al. have reported the use of an ELISA for detecting antibodies to CAstV using recombinantly expressed capsid protein [83]. Wang et al. [108] and Zhao et al. [114] also developed similar ELISA tests using recombinant capsid proteins for detecting DAstV and ANV-specific antibodies in duck and chicken serum samples.

The antigenic diversities displayed by avian astrovirus species need to be taken into consideration when using serological tests. For example, Shirai et al. [85] have demonstrated that ANV strains representing serotype 1 (ANV-1) and serotype 2 (ANV-2) exhibit very low levels of cross-reactive antibodies using indirect IF. Todd et al. [104] have recently reported the common occurrence of ANV variants, which displayed very different capsid protein sequences from those of ANV-1 and ANV-2, and which, on this basis, are likely to be antigenically different. This view is supported by a recent work by de Wit et al. [16], who showed that serum samples containing antibodies to a novel ANV isolate (isolate 19) were not detected by an ANV-1 specific indirect IF assay. It is recognized that the CAstVs can be subdivided into two major antigenic groups, designated A and B, which display relatively low levels of cross-reactivity by indirect IF [90, 105]. For this reason, a recent CAstV seroprevalence study involved the application of two separate indirect IF tests, one based on isolate 612 for detecting antibodies to the CAstVs from Group A, and the second based on isolate 11672 for detecting antibodies to the Group B CAstVs [105]. Therefore, the antigenic diversities exhibited by ANV and CAstV need to be considered in relation to their serological diagnosis. Serological tests based on the use of single ANV or CAstV strains will not detect antibodies to all ANV or CAstV infections. This is of particular importance when testing SPF flocks to demonstrate freedom from infection.

Control

Although it has been possible to produce and maintain specific pathogen free (SPF) flocks that are free from ANV infection as determined by serological testing, the common and widespread occurrence of avian astroviruses such as ANV, CAstV, and TAstVs in commercial poultry, and, in the case of ANV, their occurrence in wild birds [45], when combined with their potential for vertical transmission, strongly suggests that their eradication from commercial poultry would not be feasible. In addition, astroviruses are very stable in the environment and resistant to inactivation by most routinely used disinfectants, which complicates its elimination from infected premises [82]. Strict biosecurity, increased down time between flocks, and effective disinfection of the premises would help reduce the likelihood of contracting avian astrovirus infections.

Convalescent antisera have been used to control outbreaks of DAstV-1 and DAstV-2 [109]. Although experimental live attenuated vaccines were also developed for these astroviruses, these were not commercially produced [109]. Similarly, commercial vaccines have not been developed for ANV, CAstV, and TAstVs. In the case of ANV and CAstV, which are known to be vertically transmitted, vaccine protocols focused on the parent flocks during rear should be considered as a method to prevent or limit transmission to progeny in ovo. In addition, breeder vaccination should ensure uniformly higher levels of maternal antibody, which would better equip the young birds to withstand early viral challenges. The development of live attenuated or inactivated avian astrovirus vaccines will both depend on the availability of cell or embryo culture methods. Alternatively, recombinant expression systems, such as recombinant baculovirus system, might be used to produce astrovirus capsid subunit vaccines. Sellers et al. [83] recently described the production of a prototype recombinant baculovirus-derived CAstV vaccine, which, following multiple administrations to breeders, offered some protection to progeny chicks against experimentally induced RSS.

Conclusions

In comparison with other avian viruses, such as fowl adenoviruses, avian paramyxoviruses, and avian coronaviruses, relatively little is known about avian astroviruses and their impact on birds. A major reason for this has been the absence in most cases of cell culture growth systems, with which viruses can be isolated and biologically characterized; however, the level of genetic diversity, species affected, and permutations of clinical diseases caused by the various members of *Astroviridae* complicate our ability to develop a cohesive narrative as other avian viruses. In spite of these challenges, within the last decade, considerable advances have been made. The different avian astroviruses are second only to the human astroviruses in terms of their level of characterization, and in many cases, such a pathogenic mechanism and immunity are the best characterized. Regardless, considerably more research is needed.

Most notably, a standardized process for naming and classifying both existing as well as new members of the genera needs to be developed. As avian virology laboratories continue to develop new detection methods, and identify and characterize novel avian astroviruses in commercial poultry as well as wild bird species, a more consistent method of comparing (molecularly or antigenically) would aid in our ability to understand the relationship of viruses within the genera as well as identifying potential virulence determinants. As part of this effort to better understand avian astroviruses, the avian virology community should be encouraged to more fully characterize representative members of each genotype, serotype, and subgroup. This would include not only sequencing, but also the development of antibodies and other reagents. Together these efforts would all be in support of increasing our understanding of the cellular, physiological, and immunological changes astroviruses induce in their various host species so that we can develop more effective prevention, control, and therapeutic strategies.

References

1. Asplin FD. Duck hepatitis: vaccination against two serological types. Vet Rec. 1965;77: 1529–30.
2. Barnes GL, Guy JS, Vaillancourt JP. Poult enteritis complex. Rev Sci Tech. 2000;19: 565–88.
3. Barnes GL, Guy JS. Poult enteritis-mortality syndrome. In: Saif YM, Barnes HJ, Glisson JR, Faddly AM, MacDougald LR, Swayne DE, editors. Diseases of poultry. 11th ed. Ames IA: Blackwell Publishing; 2003.
4. Baxendale W, Mebatsion T. The isolation and characterisation of astroviruses from chickens. Avian Pathol. 2004;33:364–70.
5. Behling-Kelly E, Schultz-Cherry S, Koci M, Kelley L, Larsen D, Brown C. Localization of astrovirus in experimentally infected turkeys as determined by in situ hybridization. Vet Pathol. 2002;39:595–8.
6. Bidin M, Lojkic I, Bidin Z, Tisljar M, Majnaric D, Mikec M. Detection and characterization of avian nephritis virus in ducklings. Avian Pathol. 2011;40:173–7.
7. Bosch A, Guix S, Krishna NK, Méndez E, Monroe SS, Pantin-Jackwood M, et al. Family Astroviridae. In: King AMQ, Adams MJ, Carstens EB, Lefkowitz EJ, editors. Virus taxonomy: ninth report of the international committee on taxonomy of viruses. London: Elsevier Inc; 2012.
8. Canelli E, Cordioli P, Barbieri I, Catella A, Pennelli D, Ceruti R, et al. Astroviruses as causative agents of poultry enteritis: genetic characterization and longitudinal studies on field conditions. Avian Dis. 2012;56(1):173–82.
9. Cattoli G, De Battisti C, Toffan A, Salviato A, Lavazza A, Cerioli M, et al. Co-circulation of distinct genetic lineages of astroviruses in turkeys and guinea fowl. Arch Virol. 2007;152:595–602.
10. Cattoli G, Toffan A, De Battisti C, Salviato A, Terregino C, Capua I. Astroviruses found in the intestinal contents of guinea fowl suffering from enteritis. Vet Rec. 2005;156:220.
11. Chu DK, Chin AW, Smith GJ, Chan KH, Guan Y, Peiris JS, et al. Detection of novel astroviruses in urban brown rats and previously known astroviruses in humans. J Gen Virol. 2010;91:2457–62.
12. Chu DK, Poon LL, Guan Y, Peiris JS. Novel astroviruses in insectivorous bats. J Virol. 2008;82:9107–14.

13. Connor TJ, McNeilly F, McFerran JB, McNulty MS. A survey of avian sera from Northern Ireland for antibody to avian nephritis virus. Avian Pathol. 1987;16:15–20.
14. Day JM, Spackman E, Pantin-Jackwood M. A multiplex RT-PCR test for the differential identification of turkey astrovirus type 1, turkey astrovirus type 2, chicken astrovirus, avian nephritis virus, and avian rotavirus. Avian Dis. 2007;51:681–4.
15. de Wit JJ, Dam GB, de Laar JM, Biermann Y, Verstegen I, Edens F, et al. Detection and characterization of a new astrovirus in chicken and turkeys with enteric and locomotion disorders. Avian Pathol. 2011;40:453–61.
16. Decaesstecker M, Meulemans. An ELISA for the detection of antibodies to avian nephritis virus and related entero-like viruses. Avian Pathol. 1991;20:523–30.
17. Decaesstecker M, Charlier G, Peeters J, Meulemans G. Pathogenicity of fowl enteroviruses. Avian Pathol. 1989;18:697–713.
18. Domanska-Blicharz K, Seroka A, Minta Z. One-year molecular survey of astrovirus infection in turkeys in Poland. Arch Virol. 2011;156:1065–72.
19. Frazier JA, Howes K, Reece RL, Kidd AW, Cavanagh D. Isolation of noncytopathic viruses implicated in the aetiology of nephritis and baby chick nephropathy and serologically related to avian nephritis virus. Avian Pathol. 1990;19:139–60.
20. Fu Y, Pan M, Wang X, Xu Y, Xie X, Knowles NJ, et al. Complete sequence of a duck astrovirus associated with fatal hepatitis in ducklings. J Gen Virol. 2009;90:1104–8.
21. Gibson CA, Chen J, Monroe SA, Denison MR. Expression and processing of nonstructural proteins of the human astroviruses. Adv Exp Med Biol. 1998;440:387–91.
22. Gough RE, Borland ED, Keymer IF, Stuart JC. An outbreak of duck hepatitis type II in commercial ducks. Avian Pathol. 1985;14:227–36.
23. Gough RE, Collins MS, Borland E, Keymer LF. Astrovirus-like particles associated with hepatitis in ducklings. Vet Rec. 1984;114:279.
24. Haider SA, Calnek BW. In-vitro isolation, propagation and characterisation of duck hepatitis virus type III. Avian Dis. 1979;23:715–29.
25. Hewson KA, O'Rourke D, Noormohammadi AH. Detection of avian nephritis virus in Australian chicken flocks. Avian Dis. 2010;54:990–3.
26. Imada T. Avian nephritis. In: Saif YM, Faddly AM, Glisson JR, MacDougald LR, Nolan LK, Swayne DE, editors. Diseases of poultry. Ames, IA: Blackwell Publishing; 2008.
27. Imada T, Maeda M, Furuta K, Yamaguchi S, Kawamura H. Pathogenicity and distribution of avian nephritis virus (G-4260 strain) in inoculated laying hens. Natl Inst Anim Health Q (Tokyo). 1983;23:43–8.
28. Imada T, Taniguchi T, Yamaguchi S, Minetoma T, Maeda M, Kawamura H. Susceptibility of chickens to avian nephritis virus at various inoculation routes and ages. Avian Dis. 1981;25:294–302.
29. Imada T, Yamaguchi S, Kawamura H. Pathogenicity for baby chicks of the G-4260 strain of the picornavirus "avian nephritis virus". Avian Dis. 1979;23:582–8.
30. Imada T, Yamaguchi S, Mase M, Tsukamoto K, Kubo M, Morooka A. Avian nephritis virus (ANV) as a new member of the family Astroviridae and construction of infectious ANV cDNA. J Virol. 2000;74:8487–93.
31. Imada T, Yamaguchi S, Miura N, Kawamura H. Antibody survey against avian nephritis virus among chickens in Japan. Natl Inst Anim Health Q (Tokyo). 1980;20:79–80.
32. Jiang B, Monroe SS, Koonin EV, Stine SE, Glass RI. RNA sequence of astrovirus: distinctive genomic organization and a putative retrovirus-like ribosomal frameshifting signal that directs the viral replicase synthesis. Proc Natl Acad Sci USA. 1993;90:10539–43.
33. Jindal N, Patnayak DP, Chander Y, Ziegler AF, Goyal SM. Comparison of capsid gene sequences of turkey astrovirus-2 from poult-enteritis-syndrome-affected and apparently healthy turkeys. Arch Virol. 2011;156:969–77.
34. Jindal N, Patnayak DP, Chander Y, Ziegler AF, Goyal SM. Detection and molecular characterization of enteric viruses from poult enteritis syndrome in turkeys. Poult Sci. 2010;89: 217–26.

35. Jindal N, Patnayak DP, Ziegler AF, Lago A, Goyal SM. Duration of growth depression and pathogen shedding in experimentally reproduced poult enteritis syndrome. Avian Dis. 2009;53:517–22.
36. Jindal N, Patnayak DP, Ziegler AF, Lago A, Goyal SM. Experimental reproduction of poult enteritis syndrome: clinical findings, growth response, and microbiology. Poult Sci. 2009;88:949–58.
37. Jindal N, Patnayak DP, Ziegler AF, Lago A, Goyal SM. A retrospective study on poult enteritis syndrome in Minnesota. Avian Dis. 2009;53:268–75.
38. Jonassen CM, Jonassen TT, Sveen TM, Grinde B. Complete genomic sequences of astroviruses from sheep and turkey: comparison with related viruses. Virus Res. 2003;91:195–201.
39. Kang KI, Icard AH, Linnemann E, Sellers HS, Mundt E. Determination of the full length sequence of a chicken astrovirus suggests a different replication mechanism. Virus Genes. 2012;44(1):45–50.
40. Koci MD, Kelley LA, Larsen D, Schultz-Cherry S. Astrovirus-induced synthesis of nitric oxide contributes to virus control during infection. J Virol. 2004;78:1564–74.
41. Koci MD, Moser LA, Kelley LA, Larsen D, Brown CC, Schultz-Cherry S. Astrovirus induces diarrhea in the absence of inflammation and cell death. J Virol. 2003;77:11798–808.
42. Koci MD, Schultz-Cherry S. Avian astroviruses. Avian Pathol. 2002;31:213–27.
43. Koci MD, Seal BS, Schultz-Cherry S. Development of an RT-PCR diagnostic test for an avian astrovirus. J Virol Methods. 2000;90:79–83.
44. Koci MD, Seal BS, Schultz-Cherry S. Molecular characterization of an avian astrovirus. J Virol. 2000;74:6173–7.
45. Kofstad T, Jonassen CM. Screening of feral and wood pigeons for viruses harbouring a conserved mobile viral element: characterization of novel astroviruses and picornaviruses. PLoS One. 2011;6:e25964.
46. Krishna NK. Identification of structural domains involved in astrovirus capsid biology. Viral Immunol. 2005;18:17–26.
47. Kurtz J, Lee T. Astrovirus gastroenteritis age distribution of antibody. Med Microbiol Immunol. 1978;166:227–30.
48. Lewis TL, Matsui SM. An astrovirus frameshift signal induces ribosomal frameshifting in vitro. Arch Virol. 1995;140:1127–35.
49. Lewis TL, Matsui SM. Astrovirus ribosomal frameshifting in an infection-transfection transient expression system. J Virol. 1996;70:2869–75.
50. Lewis TL, Matsui SM. Studies of the astrovirus signal that induces (-1) ribosomal frameshifting. Adv Exp Med Biol. 1997;412:323–30.
51. MacMicking J, Xie QW, Nathan C. Nitric oxide and macrophage function. Annu Rev Immunol. 1997;15:323–50.
52. Maeda M, Imada T, Taniguchi T, Horiuchi T. Pathological changes in chicks inoculated with the picornavirus "avian nephritis virus". Avian Dis. 1979;23:589–96.
53. Mandoki M, Bakonyi T, Ivanics E, Nemes C, Dobos-Kovacs M, Rusvai M. Phylogenetic diversity of avian nephritis virus in Hungarian chicken flocks. Avian Pathol. 2006;35:224–9.
54. Mandoki M, Dobos-Kovacs M, Bakonyi T, Rusvai M. Molecular diagnosis of avian nephritis: preliminary report. Acta Vet Hung. 2006;54:51–60.
55. Marczinke B, Bloys AJ, Brown TD, Willcocks MM, Carter MJ, Brierley I. The human astrovirus RNA-dependent RNA polymerase coding region is expressed by ribosomal frameshifting. J Virol. 1994;68:5588–95.
56. McNeilly F, Connor TJ, Calvert VM, Smyth JA, Curran WL, Morley AJ, Thompson D, Singh S, McFerran JB, Adair BM, McNulty MS. Avian Pathol. 1994 Jun;23(2):313-27.
57. McNulty MS, Curran WL, McFerran JB. Detection of astroviruses in turkey faeces by direct electron microscopy. Vet Rec. 1980;106:561.
58. McNulty MS, Connor TJ, McNeilly F, McFerran JB. Biological characterization of avian enteroviruses and enterovirus-like viruses. Avian Pathol. 1990;19:75–87.

59. McNulty MS, Connor TJ, McNeilly F. A survey of specific pathogen-free chicken flocks for antibodies to chicken anemia agent, avian nephritis virus and group A rotavirus. Avian Pathol. 1989;18:215–20.
60. Meyerhoff RR, Nighot PK, Ali RA, Blikslager AT, Koci MD. Characterization of turkey inducible nitric oxide synthase and identification of its expression in the intestinal epithelium following astrovirus infection. Comp Immunol Microbiol Infect Dis. 2012;35(1):63–9.
61. Mitchell DK. Astrovirus gastroenteritis. Pediatr Infect Dis J. 2002;21:1067–9.
62. Mockett AP, Huggins MB, Woods M, Orbell S. A compariosn of three serological methods to detect chicken and turkey antibodies to avian nephritis virus and the use of virus-specific monoclonal antibodies. Avian Pathol. 1993;22(4):751–70.
63. Monceyron C, Grinde B, Jonassen TO. Molecular characterisation of the 3′-end of the astrovirus genome. Arch Virol. 1997;142:699–706.
64. Mor SK, Abin M, Costa G, Durrani A, Jindal N, Goyal SM, et al. The role of type-2 turkey astrovirus in poult enteritis syndrome. Poult Sci. 2011;90:2747–52.
65. Moser LA, Carter M, Schultz-Cherry S. Astrovirus increases epithelial barrier permeability independently of viral replication. J Virol. 2007;81:11937–45.
66. Nicholas RAJ, Goddard RD, Luff PR. Prevalence of avian nephritis virus in England. Vet Rec. 1988;123:398.
67. Nighot PK, Moeser A, Ali RA, Blikslager AT, Koci MD. Astrovirus infection induces sodium malabsorption and redistributes sodium hydrogen exchanger expression. Virology. 2010;401:146–54.
68. Pantin-Jackwood MJ, Day JM, Jackwood MW, Spackman E. Enteric viruses detected by molecular methods in commercial chicken and turkey flocks in the United States between 2005 and 2006. Avian Dis. 2008;52:235–44.
69. Pantin-Jackwood MJ, Spackman E, Day JM. Pathogenesis of type 2 turkey astroviruses with variant capsid genes in 2-day-old specific pathogen free poults. Avian Pathol. 2008;37:193–201.
70. Pantin-Jackwood MJ, Spackman E, Day JM, Rives D. Periodic monitoring of commercial turkeys for enteric viruses indicates continuous presence of astrovirus and rotavirus on the farms. Avian Dis. 2007;51:674–80.
71. Pantin-Jackwood MJ, Spackman E, Woolcock PR. Molecular characterization and typing of chicken and turkey astroviruses circulating in the United States: implications for diagnostics. Avian Dis. 2006;50:397–404.
72. Pantin-Jackwood MJ, Spackman E, Woolcock PR. Phylogenetic analysis of turkey astroviruses reveals evidence of recombination. Virus Genes. 2006;32:187–92.
73. Pantin-Jackwood MJ, Strother KO, Mundt E, Zsak L, Day JM, Spackman E. Molecular characterization of avian astroviruses. Arch Virol. 2011;156:235–44.
74. Qureshi MA, Saif YM, Heggen-Peay CL, Edens FW, Havenstein GB. Induction of functional defects in macrophages by a poult enteritis and mortality syndrome-associated turkey astrovirus. Avian Dis. 2001;45:853–61.
75. Qureshi MA, Yu M, Saif YM. A novel "small round virus" inducing poult enteritis and mortality syndrome and associated immune alterations. Avian Dis. 2000;44:275–83.
76. Reynolds DL, Saif YM. Astrovirus: a cause of an enteric disease in turkey poults. Avian Dis. 1986;30:728–35.
77. Reynolds DL, Saif YM, Theil KW. Enteric viral infections of turkey poults: incidence of infection. Avian Dis. 1987;31:272–6.
78. Reynolds DL, Saif YM, Theil KW. A survey of enteric viruses of turkey poults. Avian Dis. 1986;31:89–98.
79. Reynolds DL, Schultz-Cherry SL. Astrovirus infections. In: Saif YM, Faddly AM, Glisson JR, MacDougald LR, Nolan LK, Swayne DE, editors. Diseases of poultry. Ames, IA: Blackwell Publishing; 2008.
80. Saif LJ, Saif YM, Theil KW. Enteric viruses in diarrheic turkey poults. Avian Dis. 1985;29:798–811.
81. Saif YM, Saif LJ, Hofacre CL, Hayhow C, Swayne DE, Dearth RN. A small round virus associated with enteritis in turkey poults. Avian Dis. 1990;34:762–4.

82. Schultz-Cherry S, Kapczynski DR, Simmons VM, Koci MD, Brown C, Barnes HJ. Identifying agent(s) associated with poult enteritis and mortality syndrome: importance of the thymus. Avian Dis. 2000;44:256–65.
83. Schultz-Cherry S, King DJ, Koci MD. Inactivation of an astrovirus associated with poult enteritis mortality syndrome. Avian Dis. 2001;45:76–82.
84. Sellers H, Linneman E, Icard AH, Mundt E. A purified recombinant baculovirus expressed capsid protein of a new astrovirus provides partial protection to runting-stunting syndrome in chickens. Vaccine. 2010;28:1253–63.
85. Shirai J, Nakamura K, Narita M, Furuta K, Hihara H, Kawamura H. Visceral urate deposits in chicks inoculated with avian nephritis virus. Vet Rec. 1989;124:658–61.
86. Shirai J, Nakamura K, Shinohara K, Kawamura H. Pathogenicity and antigenicity of avian nephritis isolates. Avian Dis. 1991;35:49–54.
87. Shirai J, Tanimura N, Uramoto K, Narita M, Nakamura K, Kawamura H. Pathologically and serologically different avian nephritis virus isolates implicated in etiology of baby chick nephropathy. Avian Dis. 1992;36:369–77.
88. Smyth JA, Connor TJ, McNeilly F, Moffet DA, Calvert VM, McNulty MS. Studies on the pathogenicity of enterovirus-like viruses in chickens. Avian Pathol. 2007;36:119–26.
89. Smyth VJ, Jewhurst HL, Adair BM, Todd D. Detection of chicken astrovirus by reverse transcriptase-polymerase chain reaction. Avian Pathol. 2009;38:293–9.
90. Smyth VJ, Jewhurst HL, Wilkinson DS, Adair BM, Gordon AW, Todd D. Development and evaluation of real-time TaqMan(R) RT-PCR assays for the detection of avian nephritis virus and chicken astrovirus in chickens. Avian Pathol. 2010;39:467–74.
91. Smyth VJ, Todd D, Trudgett J, Lee A, Welsh MD. Capsid protein sequence diversity of chicken astrovirus. Avian Pathol. 2012;41(2):151–9.
92. Spackman D, Gough RE, Collins MS, Lanning D. Isolation of an enterovirus-like agent from the meconium of dead-in-shell chicken embryos. Vet Rec. 1984;114:216–8.
93. Spackman E, Day JM, Pantin-Jackwood MJ. Astrovirus, reovirus, and rotavirus concomitant infection causes decreased weight gain in broad-breasted white poults. Avian Dis. 2010; 54:16–21.
94. Spackman E, Kapczynski D, Sellers H. Multiplex real-time reverse transcription-polymerase chain reaction for the detection of three viruses associated with poult enteritis complex: turkey astrovirus, turkey coronavirus, and turkey reovirus. Avian Dis. 2005;49:86–91.
95. Strain E, Kelley LA, Schultz-Cherry S, Muse SV, Koci MD. Genomic analysis of closely related astroviruses. J Virol. 2008;82:5099–103.
96. Takase K, Shinohara K, Tsuneyoshi M, Yamamoto M, Yamada S. Isolation and characterization of cytopathic avian enteroviruses from broiler chickens. Avian Pathol. 1989;18:631–42.
97. Takase K, Uchimura T, Yamamoto M, Yamada S. Susceptibility of embryos and chicks, derived form immunized breeding hens, to avian nephritis virus. Avian Pathol. 1994;23: 117–25.
98. Tang Y, Ismail MM, Saif YM. Development of antigen-capture enzyme-linked immunosorbent assay and RT-PCR for detection of turkey astroviruses. Avian Dis. 2005;49:182–8.
99. Tang Y, Murgia AM, Saif YM. Molecular characterization of the capsid gene of two serotypes of turkey astroviruses. Avian Dis. 2005;49:514–9.
100. Tang Y, Murgia MV, Ward L, Saif YM. Pathogenicity of turkey astroviruses in turkey embryos and poults. Avian Dis. 2006;50:526–31.
101. Tang Y, Saif YM. Antigenicity of two turkey astrovirus isolates. Avian Dis. 2004;48: 896–901.
102. Thouvenelle ML, Haynes JS, Reynolds DL. Astrovirus infection in hatchling turkeys: histologic, morphometric, and ultrastructural findings. Avian Dis. 1995;39:328–36.
103. Todd D, Smyth VJ, Ball NW, Donnelly BM, Wylie M, Knowles NJ, et al. Identification of chicken enterovirus-like viruses, duck hepatitis virus type 2 and duck hepatitis virus type 3 as astroviruses. Avian Pathol. 2009;38:21–30.

104. Todd D, Trudgett J, McNeilly F, McBride N, Donnelly B, Smyth VJ, et al. Development and application of an RT-PCR test for detecting avian nephritis virus. Avian Pathol. 2010;39: 207–13.
105. Todd D, Trudgett J, Smyth VJ, Donnelly B, McBride N, Welsh MD. Capsid protein sequence diversity of avian nephritis virus. Avian Pathol. 2011;40:249–59.
106. Todd D, Wilkinson DS, Jewhurst HL, Wylie M, Gordon AW, Adair BM. A seroprevalence investigation of chicken astrovirus infections. Avian Pathol. 2009;38:301–9.
107. Toth TE. Studies of an agent causing mortality among ducklings immune to duck virus hepatitis. Avian Dis. 1969;13:834–46.
108. Walter JE, Briggs J, Guerrero ML, Matson DO, Pickering LK, Ruiz-Palacios G, et al. Molecular characterization of a novel recombinant strain of human astrovirus associated with gastroenteritis in children. Arch Virol. 2001;146:2357–67.
109. Wang X, Wang Y, Xie X, Zhang B, Zhang D. Expression of the C-terminal ORF2 protein of duck astrovirus for application in a serological test. J Virol Methods. 2011;171:8–12.
110. Woolcock PR. Duck hepatitis. In: Saif YM, Faddly AM, Glisson JR, MacDougald LR, Nolan LK, Swayne DE, editors. Diseases of poultry. Ames, IA: Blackwell Publishing; 2008.
111. Yamaguchi S, Imada T, Kawamura H. Characterisation of a picornavirus isolated from broiler chicks. Avian Dis. 1979;23:571–81.
112. Yu M, Ismail MM, Qureshi MA, Dearth RN, Barnes HJ, Saif YM. Viral agents associated with poult enteritis and mortality syndrome: the role of a small round virus and a turkey coronavirus. Avian Dis. 2000;44:297–304.
113. Yu M, Tang Y, Guo M, Zhang Q, Saif YM. Characterization of a small round virus associated with the poult enteritis and mortality syndrome. Avian Dis. 2000;44:600–10.
114. Zhao W, Hua XG, Yuan L, Cui L, Shan TL, Dai XQ, et al. Sequence analyses of the representative Chinese-prevalent strain of avian nephritis virus in healthy chicken flocks. Avian Dis. 2011;55:65–9.
115. Zhao W, Zhu AL, Yu Y, Yuan CL, Zhu CX, Lan DL, et al. Segmentation expression of capsid protein as an antigen for the detection of avian nephritis virus infection in chicken flocks. J Virol Methods. 2012;179(1):57–61.
116. Zhao W, Zhu AL, Yu Y, Yuan CL, Zhu CX, Yang ZB, et al. Complete sequence and genetic characterization of pigeon avian nephritis virus, a member of the family Astroviridae. Arch Virol. 2011;156:1559–65.
117. Zhao W, Zhu AL, Yuan CL, Yu Y, Zhu CX, Lan DL, et al. Detection of astrovirus infection in pigeons (Columbia livia) during an outbreak of diarrhoea. Avian Pathol. 2011;40:361–5.

Index

S. Schultz-Cherry (ed.), *Astrovirus Research: Essential Ideas, Everyday Impacts, Future Directions*, DOI 10.1007/978-1-4614-4735-1_6,

R

S

T

MIX
Papier aus verantwortungsvollen Quellen
Paper from responsible sources
FSC® C105338

If you have any concerns about our products,
you can contact us on
ProductSafety@springernature.com

In case Publisher is established outside the EU,
the EU authorized representative is:
Springer Nature Customer Service Center GmbH
Europaplatz 3, 69115 Heidelberg, Germany

Printed by Libri Plureos GmbH
in Hamburg, Germany